Rethinking Creative Cities Policy

In recent years, there has been high level of interest amongst policy-makers in the "creative city" concept, due to the anticipation of economic and social benefits from a growing cultural and creative economy. However, a lack of understanding of local social and economic contexts, as well as the complexities and challenges of cultural production, has resulted in formulaic, ineffective misguided policies.

This book is concerned, in various ways, with developing an understanding of the complex dimensions of cultural production, and with tackling the often weak and implied links between research, policy and urban planning. In particular, contributors are concerned with agents, protagonists and practices that appear to be somehow invisible to, hidden from, or indeed ignored in much contemporary creative cities policy. Drawing on case studies from the UK and the Netherlands, chapters consider creative industries and policy across a range of scales, from provincial cities and regional economies, to the global cities of London and Amsterdam.

This book was originally published as a special issue of *European Planning Studies*.

Allan Watson is a Senior Lecturer in Human Geography at Staffordshire University, Stoke-on-Trent, UK, and an Associate Director of the Globalisation and World Cities research network.

Calvin Taylor is Professor in Cultural Economy at the University of Leeds, UK, and Director of the Culture, Society and Innovation Research Hub.

Rethinking Creative Cities Policy

Invisible agents and hidden protagonists

Edited by
Allan Watson and Calvin Taylor

Routledge
Taylor & Francis Group

LONDON AND NEW YORK

First published 2016
by Routledge

2 Park Square, Milton Park, Abingdon, Oxon OX14 4RN
711 Third Avenue, New York, NY 10017, USA

Routledge is an imprint of the Taylor & Francis Group, an informa business

First issued in paperback 2017

British Library Cataloguing in Publication Data
A catalogue record for this book is available from the British Library

ISBN 13: 978-1-138-89008-4 (hbk)
ISBN 13: 978-1-138-08533-6 (pbk)

Typeset in Times New Roman
by RefineCatch Limited, Bungay, Suffolk

Publisher's Note
The publisher accepts responsibility for any inconsistencies that may have
arisen during the conversion of this book from journal articles to book chapters,
namely the possible inclusion of journal terminology.

Disclaimer
Every effort has been made to contact copyright holders for their permission to
reprint material in this book. The publishers would be grateful to hear from any
copyright holder who is not here acknowledged and will undertake to rectify
any errors or omissions in future editions of this book.

Contents

Citation Information

The chapters in this book were originally published in *European Planning Studies*, volume 22, issue 12 (December 2014). When citing this material, please use the original page numbering for each article, as follows:

Chapter 1: Introduction
Invisible Agents and Hidden Protagonists: Rethinking Creative Cities Policy
Allan Watson and Calvin Taylor
European Planning Studies, volume 22, issue 12 (December 2014) pp. 2429–2435

Chapter 2
Invisible Creativity? Highlighting the Hidden Impact of Freelancing in London's Creative Industries
Oli Mould, Tim Vorley and Kai Liu
European Planning Studies, volume 22, issue 12 (December 2014) pp. 2436–2455

Chapter 3
The Role of Universities in the Regional Creative Economies of the UK: Hidden Protagonists and the Challenge of Knowledge Transfer
Roberta Comunian, Calvin Taylor and David N. Smith
European Planning Studies, volume 22, issue 12 (December 2014) pp. 2456–2476

Chapter 4
Spatial–Relational Mapping in Socio-Institutional Perspectives of Innovation
Rachel C. Granger
European Planning Studies, volume 22, issue 12 (December 2014) pp. 2477–2489

Chapter 5
Creative City Policy and the Gap with Theory
Jan Jacob Trip and Arie Romein
European Planning Studies, volume 22, issue 12 (December 2014) pp. 2490–2509

Chapter 6

Cultural Amenities: Large and Small, Mainstream and Niche—A Conceptual Framework for Cultural Planning in an Age of Austerity
Robert C. Kloosterman
European Planning Studies, volume 22, issue 12 (December 2014) pp. 2510–2525

Please direct any queries you may have about the citations to
clsuk.permissions@cengage.com

Notes on Contributors

Roberta Comunian is a Lecturer in Cultural and Creative Industries at King's College, London, UK. Her research interests include the relationship between public and private investment in the arts, cultural regeneration and knowledge and production networks.

Rachel C. Granger is Senior Lecturer in Economics at the Business School of Middlesex University, London, UK. Her research focuses on urban economies and large metropolitan regions, and also the role of knowledge, creativity, and innovation in contemporary economies.

Robert C. Kloosterman is Professor of Economic Geography and Planning at the University of Amsterdam, The Netherlands. His research is guided by questions about how the social, economic and cultural transition of advanced urban economies that gathered pace after 1980 has affected cities and why different outcomes have emerged.

Kai Liu is a senior lecturer in economic sociology at the Faculty of Business, University of Greenwich, London, UK. His research interests include economic sociology, intercultural communication, cultural and creative industries.

Oli Mould is a Lecturer in Human Geography at Royal Holloway, University of London, UK. His book *Urban Subversion and the Creative City* was published by Routledge in 2015.

Arie Romein is a Researcher in the OTB Research Institute for the Built Environment at Delft University of Technology, Delft, The Netherlands. The focus of his work is on urban development, urban economy and urban planning policy, in particular regarding leisure activities, creative industry and knowledge-based urban development.

David N. Smith is Professor at the Centre for Research in Lifelong Learning, Glasgow Caledonian University, Glasgow, UK.

Calvin Taylor is Professor in Cultural Economy at the University of Leeds, UK, and Director of the Culture, Society and Innovation Research Hub. His research interests include the creative economy, cultural value and associated areas of innovation and entrepreneurship.

Jan Jacob Trip is a Researcher in the OTB Research Institute for the Built Environment at Delft University of Technology, Delft, The Netherlands. He conducts research on transport on special development; the focus in his work has gradually shifted towards urban development and urban economy.

Tim Vorley is a Senior Lecturer in Entrepreneurship in the Management School at the University of Sheffield, UK. An economic geographer by training, his primary research interests are in entrepreneurship, enterprise and regional economies.

Allan Watson is a Senior Lecturer in Human Geography at Staffordshire University, Stoke-on-Trent, UK, and an Associate Director of the Globalisation and World Cities research network. His research interests centre on the economic geography of the creative economy, and in particular the music industry. His book *Cultural Production in and Beyond the Recording Studio* was published by Routledge in 2014.

Invisible Agents and hidden Protagonists: Rethinking Creative Cities Policy

ALLAN WATSON* & CALVIN TAYLOR**

*Department of Geography, Staffordshire University, Stoke-on-Trent, Staffordshire, UK, **School of Performance and Cultural Industries, University of Leeds, Leeds, UK

ABSTRACT *This article acts as an introduction to the special issue on creative cities policy. We begin the article with a discussion of recent critical accounts of cultural/creative industries and creative cities policy, arguing that the failure of policies to fully understand the often hidden complexities of cultural production has fostered simplistic and often self-defeating policy design and intervention. We then move on to present a series of papers that are concerned in various ways with both developing an understanding of the complex dimensions of cultural production and with tackling the often weak and implicit links between research, policy and urban planning.*

Introduction

There has been an increased interest amongst policy-makers in recent years in the "creative city" concept. The label "creative city" refers to an approach to policy and planning that "recognises the urbanistic context and infrastructure within which creative industry innovation and growth take place" (O'Connor & Kong, 2009, p. 1), and which anticipates economic and social benefits from a growing cultural and creative economy. Such policies have been significantly influenced by Charles Landry's *The Creative City* (2000) and Richard Florida's *The Rise of the Creative Class* (2002), which considered creativity as a source of competitive advantage and positioned the creative economy firmly at the cutting edge of the post-industrial knowledge economy. Culture now appears at the heart of new ways of thinking and practicing economic development (Clammer, 2005;

1

Taylor, 2009). However, as Pratt (2010) argues, there is a fractured and loose web of justifying rationales for the creative city, and moreover complex and shifting matrices of justification and reality. Thus, he argues that "today the notion of a creative city stands as much for political and social mantra as an urban, social or economic policy, or even an aspiration" (2010, p. 14) and that "it is problematic to assume a direct correspondence between aims and objectives, policies and impacts" (2010, p. 14). The performative nature of the creative city imaginary in urban policy may offer some degree of comfort for the beleaguered urban planner but it represents a major challenge for the researcher.

One of the major academic criticisms of cultural/creative industries and creative city policy in this respect is the narrow vision of creativity often invoked in policy agendas concerned with urban and regional economic growth. For Taylor (2006) the prevalent habit of eliding cultural policy with economic development or economic objectives has inappropriately over-economized the arts and culture to the detriment of intelligent policy-making (see also Taylor, 2009). Neoliberal governmental strategies have in particular encouraged a form of urban entrepreneurialism that not only encourages places to compete for investment and certain ideal types of culturally endowed economic migrants, but also draws on a particular—and very restricted—notion of productive "creative" subjects (Luckman *et al.*, 2009). As Banks and O'Connor note, the specific contexts and dynamics within which cultural, symbolic or expressive values are produced cannot be reduced to the overarching goals of growth and profit on the traditional economic model. This, they go on to explain:

> ... produces many conflicts and confusions, especially at local level, where enthused policy makers confront a sector often sceptical or simply unable to act in the expected manner of a dynamic, emergent "growth sector". (2009, p. 368)

Understanding of local contexts, complexities and challenges of cultural production must be central to creative cities policy, given that "the same policies produce different effects and impacts under various institutional and social, cultural and economic contextual situations" (Pratt, 2010, p. 14). The lack of such an understanding at the centre of policy has resulted in "Xerox" policies that are simply copied from one place to another with no acknowledgement of different local social and economic contexts (Pratt, 2009). Thus creative city policies and agendas have become increasingly formulaic, and are often imposed on places in a damaging or unrealistic manner (Kong, 2000). The increasingly prescriptive tone of governmental strategies is reinforced when distributed by popular thinkers through networks of policy influence (Luckman *et al.*, 2009; Taylor, 2013).

The inability, unwillingness, and failure within cultural policy-making circles to understand the complex dimensions of cultural production has been a major reason for ineffective creative city policy and misguided policy instruments. Indeed for Pratt (2009), it is debatable whether a deep understanding of the creative/cultural industries has been achieved (see also Jeffcut & Pratt, 2002; Pratt, 2005; Oakley, 2006) and there remain a number of problematic relationships that are not yet fully understood, including public–private, formal–informal, and production–consumption, each of which is reaching new levels of complexity as, for example, new technologies pervade the relationships of the urban environment. As Pratt asserts, "the organisational ecology of the sector and policies necessary to support, sustain and promote it are complex, risky and unusual, much like the cultural and creative economy as a whole" (2010, p. 18). However, as Pratt suggests, developing an understanding of the complexities and challenges of cultural production carries a heavy burden of information, and insight, into the cultural and creative sector that "despite the upsurge of analyses that have occurred in the last quarter of a century, is still broadly inadequate for the burden placed upon it by an ever more enthusiastic policy and political communities" (2010, p. 18). Thus academic research has an important role to play in this respect, in providing the required information and insight to policy-makers, whilst engaging critically with new and anticipated future developments. This themed issue presents a number of papers that are each concerned, in various ways, with not only developing an understanding of the complex dimensions of cultural production, but also with tackling the often weak and implicit links between research, policy and urban planning.

Invisible Agents and Hidden Protagonists

One crucial element of the complexity of production in the cultural and creative industries is the issue of "hiddeness", that is to say the invisibility to policy, or lack of acknowledgement in policy, of particular actors and particular local creative practices that play an essential role in sustaining a healthy cultural milieu in urban centres (see Luckman et al., 2009). The first three papers in this special issue are concerned in various ways with agents, protagonists and practices that appear to be somehow invisible to, hidden from, or indeed ignored in much contemporary creative cities policy. In each paper, the identification of agents that have remained hidden from policy clearly highlights the need for more nuanced understandings of the complexities of the creative economy. In the first article Oli Mould, Tim Vorley and Kai Liu consider how the working practices of the freelance labour force in London's creative industries remain largely enigmatic to public policy. Freelancers, they argue, make up a large proportion of labour within the creative industries, sustaining creative firms, and playing a central role in a project-based work environment. Indeed the creative industries are characterized perhaps more than any other industrial sector by project-based work (Christopherson, 2004), and in some sectors, such as film and television, fragmentation and deregulation have resulted in almost universal freelance working (Saundry & Nolan, 1998; Ursell, 2000; Davenport, 2006). Despite this, the authors argue that freelancers remain a hidden driver of the creative economy, largely invisible and overlooked by public policy, a situation that is compounded by inconsistencies within both academic and public policy literature as to what constitutes a freelancer. Drawing on their case study of London, the authors argue that it is not only crucial that freelancers receive appropriate recognition in policy, but also that public policy should provide legislative support for freelance workers.

In the second paper in the issue, Roberta Comunian, Calvin Taylor, and David N. Smith argue that while regional and urban planning literature has examined the growth-promoting potential of universities very closely, their possible role in relation to regional and urban creative economic development has received less attention. Universities, the authors suggest, are "hidden protagonists" with often long and hidden associations with regional and urban creative activities. Drawing on a Triple Helix theoretical framework enables the authors to highlight the role played by public policy in the creative economy and the specific nature of the creative industries themselves, in order to understand the long-established but often informal connections with higher education. They highlight how there are important institutional and professional challenges in the possibility of Universities developing an explicit and sustainable role as new actors in regional and urban creative economies. Whilst universities have moved fairly quickly to embrace the imperatives of entrepreneurship and innovation, the art-world (Becker, 1984) relationships that span creative practice and the academy, by turns evidence institutionally opaque alternative forms of entrepreneurship and innovation, on the one hand, and, resistant strategies of subversion on the other.

In the third paper of the issue, Rachel Granger considers creative cities in terms of socio-institutional perspectives on urban innovation. The hidden agents identified here are social actors, with the authors arguing that analyses of urban innovation are "under socialised" and therefore that disconnections exists between research on urban innovation and its realities. The authors draw on a spatial-relational mapping of the arts sector in the city of Coventry, UK, to reveal important developments in urban creative economies that are of immediate concern for planners, including developments in sector and spatial convergence of creative workers, and the separation between "upperground" and "underground" activities. Their work emphasizes why recognition of the importance of social institutions, networks and relations should be central to urban planning. Thus the paper usefully extends to the literature emerging in particular from economic geography on the importance of geographical proximity for social aspects of cultural production (Pratt, 2000, 2002; Bathelt, 2002, 2005; Power & Hallencreutz, 2002; Power & Jansson, 2004; Watson, 2008).

Creative Cities Research, Policy and Planning

The discussion of hiddeness presented in the first three papers of this issue also speaks to another area of complexity with regard to creative cities policy; namely the relationship between academic research and policy and planning. In the fourth paper of the issue, Jan Jacob Trip and Arie Romein argue that while the importance of creativity and innovation for urban competitiveness has been analysed at the conceptual level in academic literature, and numerous cities have developed and implemented creative city policies in practice, the connection between policy and practice continues to be weak and implicit. Their article is concerned with narrowing the gap between theory and practice, by addressing the question of how conceptual insights into the creative city can be converted into an elaborated operational approach for local policy practice. Drawing on the example of the city of Delft in the Netherlands, they consider a three-step approach that allows theoretical insights to be applied in practice.

In the final paper of the issue, Robert Kloosterman critically examines the limited nature of current urban cultural planning, and is in particular critical of its indirect and strategic nature. Echoing our earlier call for an understanding of local contexts, complexities and challenges of cultural production to be at the centre of creative cities policy, the author argues that for cultural planning to be successful, a thorough understanding of the type of culture, the type of place and the role of local players is needed. The author is, for example, critical of the ways in which current cultural policies aimed at attracting knowledge workers and the "creative class" (Florida, 2002) tend to privilege larger cities and strengthen their agglomeration economies, at the expense of smaller cities that can offer only more stand-alone cultural amenities. Scale and location are then key in relation to policy and planning; as Kloosterman argues successful cultural planning is feasible, but only in relation to the characteristics and particularities of places.

It is notable in this context that the papers in this issue have considered creative industries and creative cities policy across a range of different scales, from the regional economies of the UK (Comunian et al., 2013), to the global cities of London (Mould et al., 2013) and Amsterdam (Kloosterman, 2013), to the smaller provincial cities of Coventry (Granger, 2013) and Delft (Trip & Romein, 2013). Following Waitt and Gibson (2009), we suggest that the papers all in particular ways highlight the importance of rethinking the creative economy *in place*; whether that be in large cities, or the small cities whose cultural production is so often ignored by policy-makers (Waitt, 2006). As Jayne et al. (2010) argue, the metropolitan focus, measurements and categories used by theorists such as Charles Landry and Richard Florida, and which have since been taken up by countless policy-makers, fail to capture the complexity and diversity of cultural production and creativity in smaller cities, or in working-class cities (see Jayne, 2004). Such limited views of the "place" of the cultural economy fail to recognize that such cities can have economically, politically and culturally vibrant cultural economies, often underpinned by scenes of vernacular creativity beyond money-making activities (see Gibson, 2012).

Conclusions

In concluding this short editorial, we suggest that the interactions between research, critique, policy-making and evaluation need to happen on a number of levels. First, we suggest that academic cultural and urban policy needs to engage with the practicalities of both cultural and urban planning, understanding its mechanisms, cycles, timeframes and priorities, and with the sedimented historical cultures of local institutions and their missions. Second, we argue that it is increasingly impossible to view urban development as a staged and managed process. Urban economies are subject to both emergent and disruptive changes which can call judgment on the success or otherwise of policy interventions. Understanding something of the emergent qualities of the creative city imaginary will enable policy-makers to avoid major mistakes and increase the likelihood of promoting appropriate initiatives. Finally, we advocate that research, knowledge production and knowledge use are an essential factor in the governance of the urban creative economy, but each needs to be understood as something of a moveable feast of provisional positions and perspectives. As such, the omniscient perspective of the creative economy cartographer needs to be complemented with the conditionality of the urban ethnographer.

References

Banks, M. & O'Connor, J. (2009) After the creative industries, *International Journal of Cultural Policy*, 15(4), pp. 365–373.

Bathelt, H. (2002) The re-emergence of a media industry cluster in Leipzig, *European Planning Studies*, 10(5), pp. 583–611.

Bathelt, H. (2005) Cluster relations in the media industry: Exploring the "distanced neighbour" paradox in Leipzig, *Regional Studies*, 39(1), pp. 105–127.

Becker, H. S. (1984) *Art Worlds* (Berkeley: University of California Press).

Christopherson, S. (2004) The divergent worlds of new media: How policy shapes work in the creative economy, *Review of Policy Research*, 21(4), pp. 543–558.

Clammer, J. (2005) Culture, development, and social theory: On cultural studies and the place of culture in development, *The Asia Pacific Journal of Anthropology*, 6(2), pp. 100–119.

Comunian, R., Taylor, C. & Smith, D. (2013) The role of universities in the regional creative economies of the UK: Hidden protagonists and the challenge of knowledge transfer, *European Planning Studies*. doi: 10.1080/09654313.2013.790589

Davenport, J. (2006) UK film companies: Project-based organisations lacking entrepreneurship and innovativeness?, *Creativity and Innovation Management*, 15(3), pp. 250–257.

Florida, R. (2002) *The Rise of the Creative Class—and How It's Transforming Work, Leisure, Community and Everyday Life* (New York: Basic Books).

Gibson, C. (2012) Cultural economy: Achievements, divergences, future prospects, *Geographical Research*, 50(3), pp. 282–290.

Granger, R. (2013) Spatial-relational mapping in socio-institutional perspectives of innovation, *European Planning Studies*. doi: 10.1080/09654313.2013.790591

Jayne, M. (2004) Culture that works? Creative industries development in a working-class city, *Capital and Class*, 28(3), pp. 199–210.

Jayne, M., Gibson, C., Waitt, G. & Bell, D. (2010) The cultural economy of small cities, *Geography Compass*, 4(9), pp. 1408–1417.

Jeffcut, P. & Pratt, A. (2002) Managing creativity in the cultural industries, *Creativity and Innovation Management*, 11(4), pp. 225–233.

Kloosterman, R. (2013) Cultural amenities: Large and small, mainstream and niche. A conceptual framework for cultural planning in an age of austerity, *European Planning Studies*. doi: 10.1080/09654313.2013.790594

Kong, L. (2000) Cultural policy in Singapore: Negotiating economic and socio- cultural agendas, *Geoforum*, 31(4), pp. 409–424.

Landry, C. (2000) *The Creative City: A Toolkit for Urban Innovators* (London: Earthscan).

Luckman, S., Gibson, C. & Lea, T. (2009) Mosquitoes in the mix: How transferable is creative city thinking?, *Singapore Journal of Tropical Geography*, 30(1), pp. 70–85.

Mould, O., Vorley T. & Liu, K. (2013) Invisible creativity? Highlighting the hidden impact of freelancing in London's creative industries, *European Planning Studies*. doi: 10.1080/09654313.2013.790587

Oakley, K. (2006) Include us out: Economic development and social policy in the creative industries, *Cultural Trends*, 15(4), pp. 255–273.

O'Connor, J. & Kong, L. (2009) Introduction, in: L. Kong & J. O'Connor (Eds) *Creative Economies, Creative Cities: Asian-European Perspectives*, pp. 1–8 (Dordrecht: Springer Press).

Power, D. & Hallencreutz, D. (2002) Profiting from creativity? The music industry in Stockholm, Sweden and Kingston, Jamaica, *Environment and Planning A*, 34(10), pp. 1833–1854.

Power, D. & Jansson, J. (2004) The emergence of a post-industrial music economy? Music and ICT synergies in Stockholm, Sweden, *Geoforum*, 35(4), pp. 425–439.

Pratt, A. C. (2000) New media, the new economy and new spaces, *Geoforum*, 31(4), pp. 425–436.

Pratt, A. C. (2002) Hot jobs in cool places. The material cultures of new media product spaces: The case of south of the market, San Francisco, *Information, Communication and Society*, 5(1), pp. 27–50.

Pratt, A. C. (2005) Cultural industries and public policy: An oxymoron? *International Journal of Cultural Policy*, 11(1), pp. 31–44.

Pratt, A. C. (2009) Policy transfer and the field of cultural and creative industries: What can be learned from Europe?, in: L. Kong & J. O'Connor (Eds) *Creative Economies, Creative Cities: Asian-European Perspectives*, pp. 9–24 (Dordrecht: Springer Press).

Pratt, A. C. (2010) Creative cities: Tensions within and between social, cultural and economic development, *City, Culture and Society*, 1(1), pp. 13–20.

Saundry, R. & Nolan, P. (1998) Regulatory change and performance in TV production, *Media, Culture & Society*, 20(3), pp. 409–426.

Taylor, C. (2006) Beyond advocacy: Developing an evidence base for regional creative industries strategies, *Cultural Trends*, 15(1), pp. 3–18.

Taylor, C. (2009) The creative industries, governance and economic development: A UK perspective, in: L. Kong & J. O'Connor (Eds) *Creative Economies, Creative Cities: Asian-European Perspectives*, pp. 153–166 (Dordrecht: Springer Press).

Taylor, C. (2013) Between culture, policy and industry: The modalities of intermediation in the creative economy. *Regional Studies*. doi: 10.1080/00343404.2012.748981.

Trip, J. J. & Romein, A. (2013) Creative city policy and the gap with theory, *European Planning Studies*. doi: 10.1080/09654313.2013.790592

Ursell, G. (2000) Television production: Issues of exploitation, commodification and subjectivity in UK television labour markets, *Media, Culture and Society*, 22(6), pp. 805–825.

Waitt, G. (2006) Creative small cities: Cityscapes, power and the arts, in: D. Bell & M. Jayne (Eds) *Small Cities: Urban Experience Beyond the Metropolis*, pp. 169–184 (London: Routledge).

Waitt, G. & Gibson, C. (2009) Creative small cities: Rethinking the creative economy in place, *Urban Studies*, 46(5&6), pp. 1223–1246.

Watson, A. (2008) Global music city: Knowledge and geographical proximity in London's recorded music industry, *Area*, 40(1), pp. 12–23.

Invisible Creativity? Highlighting the Hidden Impact of Freelancing in London's Creative Industries

OLI MOULD*, TIM VORLEY** & KAI LIU[†]

*Department of Geography, Royal Holloway, University of London, UK, **Management School, Centre for Regional Economic & Enterprise Development, University of Sheffield, UK, [†]Department for International Business and Economics, Business School, University of Greenwich, UK

ABSTRACT *The creative industries have been identified as a key sector for the UK's economic recovery. Despite the intense focus, however, the working practices of their labour force remain largely enigmatic to public policy. Particularly, freelancers, who make up a large proportion of labour within the creative industries, remain largely under-researched. This paper seeks to highlight the importance of freelancers to the creative industries through a case study of London's creative economy. Moreover, by discussing the prevalence of project-based work, this research shows there is a high propensity for firms to regularly engage with freelancers on a project basis—but it is the freelancers who often conduct the more creative aspects of the work. The paper concludes by suggesting that freelancers are a crucial component of the creative industries and should be included in future political decision-making.*

1. Introduction

The contemporary economy is widely cited to be a knowledge-based economy, although Knell and Oakley (2007, p. 2, added emphasis) note that the "creative and cultural sectors are an *apex* part of the knowledge economy". In many respects, the "creative industries" are privy to the same drivers as the knowledge economy, and given their collective importance have come to be regarded as an integral engine of economic growth and competitiveness. Indeed, the European Commission has suggested that the cultural and creative industries contribute 2.6% to EU GDP and employ in the region of 5 million workers (European Commission, 2010). Furthermore, in the UK (the geographical focus of this paper), the Department for Culture Media and Sport's (DCMS) most recent economic indicators show the creative industries contributed a not insignificant 5.6% of the UK's gross

8

value added in 2008 and comprised 4.1% of all goods and services exported (DCMS, 2010)—higher than the estimated figures for Europe as a whole. Moreover, the importance of the creative industries to the British economy has also been explicitly identified in recent reports on *Creative Britain* (2010) and *Digital Britain* (2010). These reports embody the legacy of New Labour's strategy to forge a sustainable and world-leading creative industries landscape, and even during times of fiscal austerity and under a new coalition government, the creative industries remain an important growth area in rebalancing the economy. In realizing the creative industries' position at an apex of the knowledge economy this paper considers the importance of one of the most under-researched, yet increasingly populous groups of workers in the sector—freelancers. And by providing empirical material that goes beyond a mere overview of the number of freelancers to engage with how they interact with commissioning firms in a project-based environment, this paper seeks to advance the study of freelancers across the social sciences.

Much existing public policy relating to the creative industries has tended to focus on larger established creative (media) firms, and only pays lip-service to the vast array of smaller businesses and freelancing. However, O'Connor (2007) identifies that it is these small businesses and freelancers that often sustain the larger firms, and as such represent an important, yet hidden driver of the creative industries as a sector. He goes on to suggest that the notion of the individual worker "doing it for oneself" is part of a wider cultural shift in the notion of labour away from routine and line management (O'Connor, 2007, p. 35). This *modus operandi* is highly prevalent within the creative industries, yet freelancers are largely invisible and overlooked by public policy. Consequently, the economic potential benefit of freelancing, and with them the creative industries, remains under-developed, an issue that is compounded by inconsistencies within both academic and public policy literature as to what constitutes a freelancer and the creative industries themselves (see Christophers, 2007; Higgs *et al.*, 2007; Pratt, 2008b; Vorley *et al.*, 2008). To this end, the paper elaborates the role of freelancers as central to the *entirety* of the creative industries in a "project-based" work environment (Christopherson, 2004; Grabher, 2004a, 2004b). By examining projects in this way, instead of other academic studies focused on one particular sub-sector (such as music (McRobbie, 2002), publishing (Ekinsmyth, 2002), television (Mitchell, 2005; Mould, 2008) or advertising Grabher, 2004a, 2004b), we can begin to realize the crucial role played by freelancers in performing the creative work in a firm-dominated sector, as well as stitching together the sector as a whole. Also, by adhering to a broad definition of the creative industries, we can bring crucial evidence to the political as well as academic landscape that hitherto has been nuanced and specific.

The remainder of this paper is structured in three sections. First the paper introduces the notion of freelancers in the context of the creative economy, differentiating the freelancer from other forms of economic agents and highlighting the difficulties in defining them in an economic context. Moreover, to progress the literature on the importance of freelance work, the section will make a differentiation between creative industry freelancers and other types of freelancers throughout the economy (such as self-employed labourers, plumbers or electricians, locum doctors, management consultants, etc.) through a discussion of projects as the main organizational unit for freelancers within the creative sector. Second, the paper draws on a case study of London, the epicentre of the UK's (and arguably Europe's) creative economy, to highlight how freelancing is integral in sustaining the city's creative economy. Finally, the paper reflects on the importance of freelancers and their conspicuous absence from public policy related to the creative industries, despite the fact they epitomize the very nature of today's entrepreneurial economy. The paper concludes by discussing the role that public policy can play in greater recognition and legislative support for freelancers.

2. Freelancers, Freelancing and the Creative Economy

It has long since been the case that the firm has been the *de facto* unit of economic enquiry, enjoying what Grabher (2004a, p. 105) describes as an "ontological and epistemological advantage". However, in his overview of "neoclassical" theory of the firm, Yeung (2001, p. 293) cautions against viewing the firm as "a self-contained and homogeneous "black-box" capable of producing economic outcomes". As such it would be obdurate to view the firm as either coherent or unproblematic, and uncritically privilege the "black-box" that is the firm as a single or starting point for empirical enquiry (Leyshon, 2011). Accordingly, over the past decade, the almost universal acceptance of the firm as the unit of calculation and analysis has come to be challenged within the literature (see Taylor & Asheim, 2001; Ekinsmyth, 2002; Yeung, 2001, 2006; Glucker, 2006; Jones, 2007; Weller, 2008). In contributing to this important strand of academic debate, this paper further unpacks the black-box by focusing on freelancers as the unit of analysis.

A seemingly obvious, although frequently overlooked question to begin is simply what is a freelancer? Different national territories across Europe will have different legal and fiscal definitions. And even in the UK context (thereby relevant to this paper), there are complicating phenomena relating to levels of "disguised employment" (Kitching & Smallbone, 2008). Fundamentally, however, the UK government, via HMRC (Her Majesty's Revenue & Customs), deems an individual to be freelance or self-employed if they pass the so-called "IR35" test. If an individual is employed as a client through an intermediary (a limited company for example), then the terms must be significantly different than if they were a regular employee. If there is no discernable different in contracts between an individual who is employed by a company "regularly" and someone who is contracted through an intermediary, then they are in breach of IR35. This law introduced in 2000 muddied the already murky legal and fiscal definitional waters of freelance workers in the UK.

Another way to address the political definition is to isolate the freelance occupations within the Standard Occupational Classification codes outlined by the Office for National Statistics (known a SOC codes for short). Table 1 highlights those occupations with the associated codes that have the potential for freelance employment, the major SOC grouping to which they belong and the level of qualifications characteristic of the group.

As can be seen from Table 1, there are a number of major categories where no freelance occupations existed (Sales and Customer Service, Process, Plant and Machine Operative and Elementary Occupations). In those other major categories, a certain level of qualification was required highlighting the skilled nature of freelancers.

Despite the increasingly widespread reference to freelancers within the academic and policy literature, the concept remains somewhat ill defined in these fields. In addressing this fuzziness, Kitching and Smallbone (2008, p. v) describe freelancers as "skilled professional workers who are neither employers nor employees, supplying labour on a temporary basis under a contract for services for a fee to a range of business clients". Building on this definition, we define a freelancer as an individual who works on a contractual or temporary basis offering their skills, knowledge and/or expertise to others (people, firms or governments) looking to outsource (and/or add value to) a particular labour

Table 1. SOC codes from (2010) that have the potential to contain freelance employment.

Major group	General nature of qualifications, training and experience	Occupational examples of creative freelancers
1: Managers, directors and senior officials	A significant amount of knowledge and experience of the production processes and service requirements associated with the efficient functioning of organizations and businesses	1134: Director (advertising, media, PR) 1136: Director (IT), owner (computer services) 1139: Director (research, conference organizers, corporate hospitality) 1259 Manager (theatrical productions), owner (photographic agency, radio, television, video servicing, theatrical agency)
2: Professional occupations	A degree or equivalent qualification, with some occupations requiring postgraduate qualifications and/or a formal period of experience-related training	2126: Design Engineer (computer, lighting) 2135: Analyst (business, IT, System, data), Architect (data, system, technical), Designer (Applications, IT, computer) 2136: Programmer(Software, Applications, games) 2137: Designer (web, internet, media, interactive, multimedia, new media), publisher (web, desktop) 2139: Consultant (IT, Computer, technical) 2150: Manager (research, journalism, printing and publishing) 2426: Researcher (journalism, printing, publishing, game, picture) 2431: Architect 2471: Editor (art, copy, picture, production), Journalist, writer 2472: Manager (account, media, publicity, PR); Officer (information, press, media, PR) 2473: Consultant (creative); Director (art, creative); Manager (account, advertising, appeal)

(*Continued*)

Table 1. Continued

Major group	General nature of qualifications, training and experience	Occupational examples of creative freelancers
3: Associate professional and technical occupations	An associated high-level vocational qualification, often involving a substantial period of full-time training or further study. Some additional task-related training is usually provided through a formal period of induction	3132: Analyst (technical, desk, service, support) 3218: Technician (theatre) 3235: Counsellor 3411 Abunatir, artist (fashion, press, scenic, sculptor, calligrapher, cartoonist); illustrator (book, fashion); modeller (artistic); painter (artistic, colour, landscape); restorer (art, picture) 3412: Author, biographer, editor, interpreter, linguist, novelist, speechwriter, copy writer 3413 Actor, acrobat, announcer, film artist, magician, Model, singer, choreographer 3414: Dancer, bandmaster, music arranger, composer, conductor, music copyist 3415: Musician, music organizer, performer, violinist 3416: Theatre director, film editor, film maker, stage manager, animation producer 3417: Cameraman, cinematographer, lighting designer, sound editor 3421: Artist (computer, graphic, layout, publishing), designer (art, display, exhibition, multimedia) 3422: Designer (book, bridalwear, cloth, commercial, costume, craft, furniture, games, glass, handbag, toy, textile)
4: Administrative and secretarial occupations	A good standard of general education. Certain occupations will require further additional vocational training to a well-defined standard (e.g. office skills)	5323: Contractor (painting, decorating), painter (buildings, house, industrial)
5: Skilled trades occupations	A substantial period of training, often provided by means of a work based training programme	5414: Maker (garment, dress, trouser) 6222: Beauty adviser, colour analyst, artist (body, make-up, image), tattooist, beauty therapist
6: Caring, leisure and other service occupations	A good standard of general education. Certain occupations will require further additional vocational training, often provided by means of a work-based training programme	7124: Trader (market, street) 7125: Assistant (display, merchandizing, visual), window dresser, window displayman), fashion stylist, photographic stylist, film stylist)

7: Sales and customer service occupations	A general education and a programme of work-based training related to sales procedures. Some occupations require additional specific technical knowledge but are included in this major group because the primary task involves selling	N/A
8: Process, plant and machine operatives	The knowledge and experience necessary to operate vehicles and other mobile and stationary machinery, to operate and monitor industrial plant and equipment, to assemble products from component parts according to strict rules and procedures and subject assembled parts to routine tests. Most occupations in this major group will specify a minimum standard of competence for associated tasks and will have a related period of formal training.	N/A
9: Elementary occupations	Occupations classified at this level will usually require a minimum general level of education (that is, that which is acquired by the end of the period of compulsory education). Some occupations at this level will also have short periods of work-related training in areas such as health and safety, food hygiene and customer service requirements	N/A

cost. Such a definition resonates with McRobbie (2002, p. 519) who suggests that "the individual [freelancer] becomes his or her own enterprise, sometimes presiding over two separate companies at the one-time". As such, we contend that there remains a need to differentiate between the freelancer and the firm.

The dawn of an enterprise culture in the UK that prevailed under Thatcherism in the 1980s has culminated in an "ideological melange" that has recast capitalism with the individual at its core. This individualism brings with it an acute recognition of "the self" (McRobbie, 2002), and Storey *et al.* (2005) suggest that freelancers personify what they term the "enterprising self". By taking initiative and responsibility for (their own) economic production in this sense, freelancers can be seen as the embodiment of the entrepreneurial or enterprise society. Over the past three decades, freelancers have proliferated in all spheres of the economy, exhibiting entrepreneurial talent which is free from "corporate control" (Entwistle & Wissinger, 2006). Moreover, in the present climate of economic uncertainty, freelancers, like entrepreneurs, are able to adapt and respond thereby making them integral to the economic recovery.

Elsewhere, however, the importance of the freelancer has been met with greater scepticism. Indeed, Faulkner (1983) contends that far from having the freedom to choose, freelancers commonly have no choice and freelancing represents the only way to enter the labour market. In this vein, Barley and Kunda (2004, p. 187) suggest that "like resident aliens, contractors coexisted with full-time employees without being full-fledged members of the community; rather, they were viewed as outsiders, who participated in the work but did not belong". In this respect, freelancers can be seen to have a *role* but not a *place*. Consequently, the nature of freelance work is complex. By its nature, freelance work is characterized by employment for a fixed period of time or, in the case of this paper, for a particular project. Consequently, Storey *et al.* (2005, p. 1040) find that "termination is an intrinsic property of the freelance employment ... [and] that responsibility for a continuous stream of work and income lay with the freelancer". On this basis, Mitchell (2005, pp. 1–2) concludes that being a freelancer represents little more than "thinly disguised unemployment", and is testament to the fact that freelance work can be challenging and difficult to sustain. Moreover, Stanworth and Stanworth (1997) contend that freelancers pose a considerable threat to in-house labour, and since they do not command the same benefits, perquisites and security or the same costs, they represent an attractive option to employers. Together, these issues mean freelancers can have a disruptive impact upon the norms and conventions of "traditional" work and employment within the firm environment. A descriptive analysis of freelancers is also offered by Nies and Pedersini (2003), who distinguish between the "false" and "forced" and "true" characteristics of freelancers. According to this distinction, "false freelancers" are workers who are essentially employees but falsely register as self-employed, whereas "forced freelancers" are those working outside of the organizational structure of the firm, although are effectively dependent on a single employer. "True freelancers" are those individuals who operate more closely with the descriptions of freelancers offered above, i.e. those who go into freelance work for legitimate economic reasons (such as love of the work, independence, niche markets, etc.).

Thus, while the term "freelancer" is not unproblematic it has now come to permeate the everyday vocabularies or work, referring to the individual who operated as an independent economic unit by responding to opportunities of the market in which they work. In particular, freelancers are characterized by their flexibility, typically working on short-term projects and at the behest of the (possibly multiple) contractor(s). The fluidity and the staccato nature of freelancing means that it is often difficult to ascertain the full extent of freelance work, hence the relative invisibility of freelancers in official productivity and growth statistics. The distinct lack of empirical data about the number of freelancers explains the lack of governmental support, particularly in the creative industries where they are so prevalent.

Accordingly, it is difficult quantify the number of freelancers, not least due to inconsistencies both in defining and in identifying freelancers. Figures for the freelancer population in Europe are elusive as they are often grouped in with "micro-businesses". But in the UK, according to Kitchings and Smallbone (2012), the number of 'freelancers' has increased from 1.39 to 1.56 million during the period 2008 to 2011, a rise of 12%. With respect to the creative industries in particular, the European Commission for Culture suggests that 80% of enterprises in the cultural and creative industries are SMEs, with 60% of micro-enterprises consisting of one to three employees. There is also evidence to suggest that workers in the creative sector are twice as likely to be self-employed than elsewhere in the economy (European Commission, 2011). In the UK, Skillset, the Sector Skills Council, has published a series of sector profiles including media, film and television among others. The most recent survey, conducted in 2009, found total employment to be approximately 500,000, with almost 25% of workers to be freelancers (Skillset, 2011). Within the creative industries, there is high occupational variance among freelancers, peaking at 72% in "camera/photography" while occupations involved with sales and distribution have tended to retain more traditional models of employment with no freelancers. Elsewhere the New Deal of the Mind (NDotM), a coalition representing Britain's creative talent, found the creative industries to have "a far higher percentage of freelance workers—around 40% of the total, compared with 12% in the economy as a whole" (NDotM, 2011, p. 1). These finding are further supported by the UK Design Council (2010) who identified the number of freelancers to have increased by almost 40% since 2005, with more than a quarter of designers in the UK now working as freelancers. Elsewhere, Gunnell and Bright (2010) find that 70% of those regularly working in organizations funded by the Arts Council are employed on a freelance basis. Evidence from other (sub-)sector councils, including the recently abolished UK Film Council, also serve to highlight the importance of freelancers for the growth of the creative industries.

Despite these apparent discrepancies concerning the size of the freelancer population, there is a consensus in the literature on the creative economy regarding their importance (see Caves, 2000; McRobbie, 2002; O'Connor, 2007; Tempest, 2009). Having provided a quantitative insight into the relative size and growth of freelancers and freelancing in the creative industries, there is little research focusing on how freelancers work, or indeed if there are differences in working practices to freelancers in other areas of work. Academic research on the creative industries is fragmented across many different disciplines in the social sciences and encompasses multiple empirical agendas which tend to focus on a specific sub-sector of the creative industries (such as film or music industries) rather than working practices *per se*. Interestingly, although the creative industries are represented by a seemingly disparate array of sub-sectors, Christophers (2007) and Pratt (2008a) identify such divisions as unnecessary and artificial. Moreover, such divisions run counter to the inherent tendency of creative industry workers to collaborate and often organize themselves into temporary networks of projects that disband on completion.

Hence, this paper will focus on the creative industries as a whole to empirically account for such cross-over, a phenomenon which is unaccountable for specifically sectoral studies that has dominated the literature thus far.

The creative industries have come to be characterized by this "project-based" or "projectized" nature of work (Christopherson, 2004; Grabher, 2004a), whereby freelancers and firms work together to complete a project (i.e. film, television product, computer game, exhibition, etc.) over a specific period of time. The nature of project based work, and the tacit learning that is inherent within it, creates an environment of uncertainty and precariousness for creative industry freelancers. Grabher (2004a, 2004b) has been a key protagonist of the project within the academic literature, notably championing what he terms "project ecologies". Assuming a "non-essentialist perspective", Grabher views projects as dependent on time and place, an approach that entails "thinking about [project] knowledge spaces *topologically*" (Grabher, 2004a, p. 106, original emphasis). With reference to the advertising industry in London and the software industry in Munich, he elaborates the non-linear and networked connectivity of project-based industries. Grabher's project ecologies provides a framework for how the (creative industry) firm organizes itself within "knowledge spaces" and is situated in a wider "epistemic community". Viewing projects as based in communities of commensurable, but tacit knowledge, collaborative learning and work sees knowledge passing from agent to agent. To this end, Grabher (2004a, p. 111) identifies how "agents are bound together by their commitment to enhance a specific type of knowledge ... and evaluated in the contribution of the agent to the cognitive goal with regard to the criteria set by the procedural authority". Therefore, for freelancers to succeed in the creative industries, they need to be part of these "knowledge spaces". Moreover, they need to undertake intense networking practices in order to keep "up-to-date" with the current trends and fashions (something which, according to Grabher (2004a), is key to success in London's advertising industry) as well work together successfully with other freelancers and economic agents.

As a model of organization, the proliferation of project-based work, has culminated in the progressive meshing of the sub-sectors of the creative industries and resulted in an almost sector-wide rise in freelance working (Davenport, 2006). Moreover, the projectization of modern industrial practices poses a series of questions which "require researchers to study in detail the nature, processes and requirements of project formation and the interactions between project members, not simply for the duration of the project, but after the project comes to an end" (Ekinsmyth, 2002, p. 231). There are negative connotations to project work too. Entwistle and Wissinger (2006, p. 782) describe how its "unpredictable, erratic and precarious [nature] makes considerable demands upon the individual in terms of their self-reliance and resourcefulness". Consequently, rather than focusing on the project as an singular economic unit, there is greater value in considering the *long-durée* of project-based networking, whereby projects represent a particular form, or phase, of a short-lived project-based network, where the "lifespan will be *limited to the duration of the project*" (Rifkin, 2005, p. 363), but for part of what Grabher (2004a) calls the "project ecology". And it is this networking process that is particularly evident among freelancers involved in the creative industries, more so than any other sector or industry.

So it is here that we see the fundamental difference between creative economy freelancers and those populating other areas of the economy. It is the success of the project network that will ultimately determine the success (and hence future employability) of the freelancer, and their ability to navigate a wider community of "project ecologies" (Grabher, 2004a). Freelance workers in other areas of the economy will be employed as an individual for a commissioning agent (a firm, a client or a government body) and work the majority of the contract as an individual. Locum doctors, for example, while having the flexibility in working hours, it comes at a cost; "poor professional status [and] managing oneself as a self-employed professional with no managerial support" (Fieldhouse, 2007, p. 33). So the individuality of work is prominent. Freelance management consultants, for example, may be hired by firms to work *for* a team (Malone & Laubacher, 1998), but their success is not dependent upon the successful completion of a product as is the case in the creative economy (although it may be performance related). Similar levels of individuality can be found in the more "traditional" self-employed occupations of electricians, plumbers, carpenters, etc. There is an argument that the construction industry has similar project-based characteristics of institutionalized learning through projects (Kamara *et al.*, 2002). However, within this sector, there is a dominance of larger firms rather than freelancers, and while the project-based nature is key, it is the common characteristic of the large construction firms involved with arguably lower levels of risk.

Also, in the creative industries, the prevalence of projects as the basis for organizing work can be seen to impact more acutely on the freelancers than those workers in established companies. For example, companies with multiple workers can split their time and resources between projects where there is a clash, whereas given the time-specific nature of event, generally a freelancer in the same situation would be forced to choose between working on one project over another (which are often place-specific). There is also a lack of unionization in all but the most coordinated sectors of the creative industries (such as the US film industry—the power of the union exemplified by the effects of the Writers Guild of America strike in 2007/2008) further exacerbates the precariousness and vulnerability of freelance workers (Heery *et al.*, 2004; Saundry *et al.*, 2006).

Therefore, by showing how creative industry freelancers operate qualitatively differently to other types of freelancers (i.e. through the premise of project-based networks), how can we begin to exemplify this through empirical work? Is there a way of describing their contribution to the project-based nature of cultural production? In order to develop a more nuanced understanding about the organizational implications of freelancing as an emergent working practice and the operational practice of freelancers within project-based environments, it is necessary to disentangle project ecologies. Moreover, this will serve to highlight the paradigmatically different role performed by freelancers compared to firms. The role of freelancing in the more "creative" aspects of cultural production has been a long-held belief in the literature (O'Connor, 2007; Mould, 2008; Pratt, 2008a), but without the empirical weight to support it. Also, research on freelancers (much of which has been discussed above) has used qualitative research techniques on the freelancers themselves to give accounts of working practices and is done so within the premise of particular sub-sectors. This is vital in highlighting the value of freelance work, but this paper seeks not to simply repeat these data on a larger industry-wide scale, but to add hitherto hidden variance to the body of data by focusing on the "other side" of the conversation, namely the role of the commissioning agent who hires the freelancers initially. In so doing, the data give a broader statistical view of where freelancing is situated in terms of its productivity and creative input in the project environment.

Researching Freelancing in London's Creative Economy

London is the epicentre of Europe's creative industry activity, and accounts for 32% of jobs in the creative industries in the UK (Skillset, 2011). The intense interactivity and networking that goes on there creates opportunities to study the complexity of freelancers' project networks. The methodology for this paper was part of wider project and was in the form of a survey that was sent out to a representative sample of London's creative industry landscape during a time period spanning six months (February 2008 to July 2008). Given the problematic definition of the creative industries as a single entity in academic and political debate (see Christophers, 2007), a decision was made to utilize a broad concept of a "creative industry" company as possible, to capture a wider range of businesses. A broad range of sub-sectors was included, from film and television, architecture advertising, publishing and musicians. Two thousand companies were contacted and 673 returns were collected (a response rate of 33.6%). Of those collected, 47 responses indicated that the company employs one full-time person, indicating that they are an operational freelance company. When asked about their legal status, 59 responded as freelance, sole trader or self-employed. With a crossover of 24 (i.e. they indicated there was one full-time employee and identified themselves as a sole-trader, self-employed or freelancer), we identified 82 respondents that have a high degree of freelance qualities, traditionally defined—12% of the total, which is lower than the estimates in previous public policy literatures. Of the 673, 301 (44.7%) responded saying that they *did not* employ any freelancers in the previous 12 months.

This was a critical division within the data, as certain characteristics of creative industry companies could be compared, to highlight the differences that freelancers make to the output and characteristics of that firm. Table 2 gives an indication of the size of the firm (in terms of employees) that did or did not employ freelancers. As can be seen, companies that did employ freelancers in 2007 had an average of 55.21 full-time staff, much larger that the 10.72 full-time staff in companies that did not employ any freelancers.

This suggests that larger firms are more likely to hire freelancers, perhaps an indirect indication that they have the funds and the workloads available to do so. These data by themselves do not tell us much about the nature of freelance work, only that there is a tendency for larger companies to hire more freelancers. However, what role the freelancers undertake in relation to the firm is an important distinction to make. To analyse this trend, a further question was asked about the nature of the companies' creative activity. Capturing and articulating the creative process is a difficult task due to the inherent intangibility of the subject matter, but in order to ascertain the "level" or type of creativity that is being undertaken by these businesses and the associated freelancers, a typology of creative practice was formulated. The first "type" of creativity is the "development of brand new

Table 2. Comparison of companies that DO and DO NOT hire freelancers

Number of employers who are	Companies that DO NOT hire freelancers	Companies that DO hire freelancers
Full time	10.72	55.21
Part-time	0.6	4.02
Volunteer	3	3.34

products/services". The second was 'development and extension of existing products/services. The third was "upkeep of existing products/services that have remained unchanged". And the fourth was "replicating products/services already in the market". These four processes (Brand New, Development, Upkeep, Replication) are qualitative representations of the "levels" of creative activity, with Brand New relating to pure innovation and Replication denoting the "least" creative practice. Of course, the creative process is far from being as clear-cut and linear as this typology suggests, but in order to gauge how much creative input London's creative industry freelancers have, then it was necessary to provide some objectivity and delineation. Care was also taken not to bias a particular category and for one not to be seen as more creative or more appealing than another (as, after all, who would want to be seen as uncreative? (Pratt, 2008b)).

Each respondent to the survey was asked to give an approximate percentage of their activity that fell into each of those categories. Table 3 shows the averages of those percentages grouped into those companies that did hire freelancers in 2008, those that did not and the freelance companies. As can be seen, those companies that hire freelancers do more of the "Brand New" type of creative work. Those companies that do not hire freelancers specialized more in the Upkeep of existing goods and services. And as for the freelance companies themselves, they are more inclined to carry out the Brand New creative work. All this suggests that in London at least freelancers are more likely to carry out the more creative parts of the production process, namely the development of brand new goods and services. While the nuance of the creative work is unexplored, there remains a significant statistical bias towards the fact that freelancers carry out the more creative aspects of production, a finding that highlights the importance of freelancers to the creative economy as a whole.

Another variable form the data relevant to the relationship between the firm and the freelancer is the type of "contracting" used. A question was asked to the respondents as to the importance placed on formal contracts placed on certain external economic relations, one of which being the hiring of freelancers. The respondents were given the opportunity to say whether they used formal arrangements (i.e. advertising through agencies and the use of legal contracts binding in workers prior to employment) or informal arrangements (gaining work through verbal, tacit agreements and social networking) in recruiting freelancers. Of the 673 respondents, 483 (72%) said they used formal legal contracts. In terms of the relative size of the companies in relation to this variable, of the 483 who said they used formal contracts, the average number of employees was 45.5. Of the 190 who said they used more informal agreements, the average number of

Table 3. Comparison of the type of creative work

Type of creative work	Average % of organizational resources of companies that DO hire freelancers	Average % of organizational resources of companies that DO NOT hire freelancers	Average % of organizational resources of FREELANCERS
Brand New	36.1	33.3	38.6
Development	31.6	29.6	30.4
Upkeep	35.2	38.8	36.6
Replication	14.9	10.2	11.3

people employed in the business was 6.2. And of the 82 identified freelance companies, all but 10 said they used informal agreements in securing the work of other freelancers. This therefore suggests that the smaller the company the less likely they are to use formal contracts in obtaining the people needed to do the job required.

In the previous section, we articulated the qualitative difference between freelancers within the creative economy and other contemporary sectors. The presence of "project ecologies" (Grabher, 2004a) is a key characteristic in this regard, and so, to further develop the empirical findings, it is pertinent to see if these freelancers operate in a project-driven environment. In the survey, among a series of questions about operational behaviour, a prominent question was asked about projects: "Do you consider yourself a project-based organization?" A definition was offered which read, "Project-based organizations allocate their resources and structure their operations according to the temporary project(s) they have—they may undertake one project at a time or have a portfolio of projects". Out of all the responses, 93 (13.8%) said that they *did not* consider themselves to be a project-based organization. The rest of the companies that said they were a project-based organization and were then asked to give a description of their most recent project. A word cloud of all the responses can be seen in Figure 1. This involves taking the whole response by each respondent and cleaning them for errors and duplication. The word cloud represents the most used words within all responses by size.

As can be seen, the major "descriptors" used are "design", "production" and (unsurprisingly) "project". Some of the other important words used by the companies surveyed include "planning", "film", "development", "television", "website" and "event". This qualitative textual analysis shows how it is design and visual art (film, television and websites) dominate the themes of projects. Also, the process descriptors such as "planning", "production" and "development" indicate that these projects are very much focused on the creation of new material. Less prominent are the words such as "distribution", "signage" and "branding".

A selection of the answers from the same question but from just the freelance companies included:

Figure 1. Word cloud of text from all responses to 'last project' question.

- consultation on fashion design projects
- design and layout for a newsletter for print and/or email and possibly related information built into a website
- commission for a ring from a private client
- lighting design for events, parties, theatre
- installation of all theatre specific services for sub-1000 seat theatres
- portrait photographs for a band for their website
- training video for bank on anti-money laundering procedures

As can be seen from these sampled responses, some of the activities can be seen as key creative processes (e.g. fashion consultation, design, portrait photography) with some less so (e.g. installation). Not all respondents adequately answered this question, however, it is clear from the multitude of responses from freelance companies (numbering 82) that they engage with the creative process, and less so with distribution, marketing, sale or administrative tasks.

For more quantitative analysis, the companies that answered yes and no to whether they engaged in projects were then split into two groups and the average number and type of staff they hired is plotted in Figure 2. As can be seen in this graph, project-based organizations hire more full-time staff, part-time and crucially more freelancers. While the ratio of freelancers to full-time staff is relatively similar (slightly less for non-project-based organizations), they are used more readily than part-time or volunteer staff in project-based organizations. These data then suggest that firms operating primarily in projects have larger staff teams, and more likely to hire freelancers, resonating with the data in Figure 2.

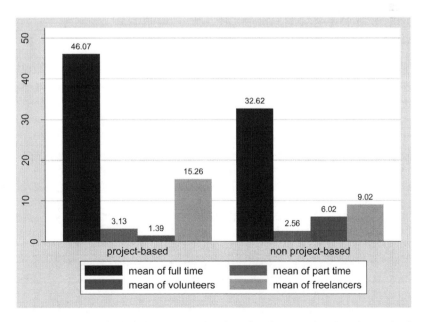

Figure 2. Labour comparison between project-based and non-project-based organizations.

So freelancers tend to be contracted by larger more structured firms to carry out the more creative parts of their operation. Smaller firms, in terms of full-time employees, are less likely to do so. Part of this may result from the fact that particular projects, such as a motion picture, television programme, theatre production, a computer game or even construction of a building requires a large number of different skilled individuals. The operating and administrative duties may also be more frequent hence the need for large numbers of in-house full-time staff. And it is the larger firms that tend to be involved in these projects, mainly by funding them. Within the survey, some of the largest firms (in terms of number of staff and turnover) to respond to the survey were London-based television and feature film companies and while they are involved in multiple projects hiring a myriad of freelancers (from runners to directors), they also have a large administrative team that deals with the financial, legal and bureaucratic nature of television and film-making. And more often than not, due to the intense social networking of freelancers, they will need to be hired as a team hence further increasing the costs. This is in contrast to, for example, some publishing companies who will hire a single photographer for a small fee to provide an image to accompany an article or the front cover of a publication.

London is widely acknowledged as the UK's hub of the creative industries (Skillset, 2011), and this study has shown freelancers to be an integral part of the creative industry landscape. Indeed, far from being thinly veiled unemployment this paper has shown the prevalence of freelancers as individuals who provide their skills, knowledge and/or expertise on a project-by-project basis. Freelancers, therefore, can be understood as significant to the creative process and integral to the creative industries as a whole. This study represents a foray into researching the prevalence of freelancers in the creative industries and identifies the need for further research to explore freelancing and freelancers in more depth.

3. Invisible But Not Insignificant

Having presented a portrait of freelancing in London it is apparent that the work and working practice of freelancers remain largely invisible/elusive in terms of how they are represented in official statistics and understood in terms of the literature on labour markets and work. The empirical study has illuminated the creative nature of the work undertaken by freelancers and has begun to show that freelancing is anything but insignificant to the creative industries. Indeed, this paper argues that freelancing has fast become a central feature of creative economic landscape in London, and across the UK economy more generally. In order to provide an alternative rhetoric to the existing data on creative industry freelance work already undertaken within the social sciences, this paper has deliberately taken a sector-wide, firm orientated approach. Therefore, it is beyond the scope of this paper to consider the intricacies of freelancing and freelance work. Therefore, we briefly reflect on the form, function and future of freelancers working in the creative industries and prospects for additional research, and how we can forward the case for further incorporation of freelancers into a political economic narrative.

In terms of the form of freelancers described, it is clear that within the creative industries there are an increasing number of freelancers operating with established firms on a regular basis. With little data on the working practice of freelancer in terms of number of clients/projects or their work status, it is difficult to differentiate between Nies and Pedersini's (2003) typology of freelancers. That said, the data suggest that all most all of the instances of freelancing identified were what Nies and Pedersini refer to as "true" freelancers—that is to say individuals engaging in a variety of projects for a variety of clients. It is much more difficult to distinguish between what Nies and Pedersini refer to "false" and "forced" freelancers, i.e. those who are work for a single client, effectively as an employee, either under their own volition or that of the employer. With the majority of the firms surveyed stating that they either recruited freelancers on recommendation/ reputation or through a third-party agency on a project-by-project basis, which would support the view that there are few instances of "false" and "forced" freelancing. With the dominant form of freelance work occurring on a project-by-project basis, this supports Grabher's (2004b) view of a projectized economy in which the social network(ing) and reputation of freelancers is critical to their performance as competitive economic agents.

From the perspective of the firm, the survey identified the function of freelancers as providing an important creative input into projects managed by larger firms working in the creative industries. As highlighted in Table 3, freelancers were primarily utilized to undertake what was categorized as "Brand New" work rather than simply providing extra capacity support to existing functions, i.e. "Upkeep". While the survey highlights that freelancers are engaged in a diverse range of activities, to further discern the full extent of the functions that freelancers fulfil requires further research. There is a need for both quantitative and qualitative research to better understand the scope and remit of freelancers in order to assess what they contribute to specific projects and the creative industries more generally. While the instances of freelancing identified in this paper are consistent with Storey et al.'s (2005) notion of the "enterprising self", further research is required understand variations in the working practices and functions of freelancers.

Freelancers have come to represent an important feature of the landscape in the creative industries. Indeed, the growth of freelancing can, at least in part, be considered a reason for the continued competitiveness of the UK's creative industries in the wake of the 2007 financial crisis. Rather than a simply a phase, it appears inevitable that freelancing has become a core mode of working and with it permanent feature of the creative industries. Freelancers have been shown in this paper as crucial to the creative industries, and we contend that the future is set to see freelancers become more visible and more significant as a result of the "project-based turn" in how economic activities are organized. Indeed, there is also a crossover here between debates on freelancers and entrepreneurship (see Bilton, 2007), since freelancing can be understood as an entrepreneurial endeavour. Given the prominence of entrepreneurship in public policy, both nationally and internationally, as sources of innovation, competitiveness and growth, freelancing can be seen to play an important role both in London's creative industries and the sector more generally across Europe. Therefore, if the potential of freelancing is to be taken advantage of and stimulated even further, then there is a need for public policy to better recognize and support them, in much the same way that SMEs have been supported as a subset of European industrial policy.

Having outlined the form, function and future of freelancing in the creative industries, we have elaborated on the foundations from the survey, as what is much needed further research into freelancing and freelancers in the creative industries. Given the diverse array of industrial activity, and increasingly the cross-fertilization of sectors, which comprises the creative industries, there is no such thing as the "typical" freelancer, not least, as there is no such thing as a "typical" project. That said, while heterogeneous in nature the survey indicates that there are common characteristics in the form and function of freelancing, with respect to how it tends to add creative value to projects. Further research needs to give particular attention to the value added by freelancing, and how they differ from other economic actors. Only then will a more comprehensive picture emerge as to the economic health and viability of the sector.

4. Conclusion

This paper has furthered the understanding of freelance work by showing the significance of freelancers working in the creative industries in London and highlighted what they bring to the creative process of project-based work. By exploring the sector as a whole (notwithstanding the complicated definitional issues) and not as specific sub-sectors, this paper has brought empirical and statistical evidence to an inherently patchy and "messy" field of economic policy. Furthermore, by focusing on firms *and* freelancers, the data of this paper expand the information on project networks that has hitherto remained largely assumed within social science literature on freelancers. In exploring the creative industries landscape it quickly becomes evident that freelancing as a working practice has become an extremely prevalent and an increasingly important mode of work. Therefore, the form and function of freelancers can be regarded as complimentary to the prevailing industrial structure rather than in competition with it. If freelancers are to fulfil their potential and continue to act as a source of competitiveness then public policy needs to be better related to the needs of freelancers in much the same way it has come to support SMEs. Consequently, recognizing and supporting freelancers in UK public is essential for sustaining the growth of the creative industries. In addition to presenting a portrait of freelancing as a part of London's creative industries and identifying the absence of freelancers from public policy debates, this paper has signposted numerous avenues for further research relating to the form, function and future of freelancers.

Overall the prevalence of freelancers needs to be understood in terms of the organizational dynamics of the creative industries as a whole. The pervasive nature of project-based work in London's creative industries was highlighted by 86.2% of firms surveyed, who operated on a project-by-project basis. The finding of the study indicated that there was a tendency for firms to employ the services of freelancers to add value to the creative process, while in-house staff undertook a range of core functions across different project. The projectization of work can be seen as a significant catalyst to the prevalence of freelancing and resulted in freelancers specializing in their respective aspects of creative production. While identifying these trends among freelancers, as noted above, the nuances remained largely unexplored; therefore, further research needs to focus on the complex relationship of between freelancers and creative industries.

More broadly, this paper illustrates how freelancers encounter difficulties in the form of increased risk, short-termism, low levels of unionized representation and constant termination. It must be stressed, however, that we are acutely aware that despite these difficulties, which varies radically across sectors and geographically, there are many freelancers who do not regard such risk as "negative". The liberation from corporate control is an important factor (Entwistle & Wissinger, 2006), or simply the fact that they are doing what they love means that many individual workers are contented with the lower wages, higher risks of unemployment and may not be affected by the other institutional difficulties outlined in this paper. However, this should not mean that those freelancers and individual workers who do face difficulties should not be championed, and policy should not be questioned.

As was argued in the first part of this paper, freelancers operate differently in the creative economy to other industrial sectors. The major proponent of that difference is the proliferation of project-based work. As a result, any future political decision-making with regard the creative industries needs to be even more acutely aware of the significance of freelancers in the so-called "project ecologies", as they continue to increase their share of the formulation of creative content in lieu of traditional firms. In the current UK coalition government, there are a number of policies designed to increase the productivity of the creative industries sector, but in order to be more effective, then these policies will need an evidence base of the importance of freelancers to project networks—an evidence base which this paper has presented in the first instance, and lays the groundwork for future applied study. Within the European policy, the cultural and creative industries are seen as a key growth area, and given that workers in this sector are twice as likely to be self-employed than elsewhere in the economy (European Commission, 2011), the marginalization of freelancers in economic political decision-making needs to be addressed. Many of the creative sectors that are being championed by the current UK Con-Lib government (such as the computer game industry and the film industry) are heavily reliant upon freelance workers and operate increasingly around projects. The promotion of the individual as key to economic prosperity was further highlighted as a key tenant of the current UK government's policy when Prime Minister David Cameron spoke in March 2011. He outlined the importance of entrepreneurs and how it was crucial to "back small firms" (BBC, 2011) in attempts to stimulate the UK economy to recover from recession. Freelancers then will potentially play a major role in this entrepreneurial landscape, particularly in the creative industries. Therefore, greater recognition and more research about the working practices of freelancers, the economic benefits of freelancing and how their creativity can be harnessed is vital if we are to harness the full potential of the creative industries.

References

Barley, S. & Kunda, G. (2004) *Gurus, Hired Guns, and Warm Bodies* (NJ: Princeton University Press).

BBC (2011) David Cameron says enterprise is only hope for growth. Available at http://www.bbc.co.uk/news/uk-politics-12657524 (accessed 9 March 2011).

Bilton, C. (2007) *Management and Creativity: From Creative Industries to Creative Management* (London: John Wiley & Sons).

Caves, R. (2000) *Creative Industries* (Cambridge, MA: Harvard University Press).

Christophers, B. (2007) Enframing creativity: Power, geographical knowledges and the media economy, *Transactions of the Institute of British Geographers*, 32(2), pp. 235–247.

Christopherson, S. (2004) The divergent worlds of new media: How policy shapes work in the creative economy, *Review of Policy Research*, 21(4), pp. 543–558.

Davenport, J. (2006) UK film companies: Project-based organizations lacking entrepreneurship and innovativeness? *Creativity and Innovation Management*, 15(3), pp. 250–257.

DCMS (2010) *Creative Industries Economic Estimates* (London: HMSO).

Design Council (2010) *UK Design Industry*. Available at http://www.designcouncil.org.uk/Documents/Documents/Publications/Research/DesignIndustryResearch2010/DesignIndustryResearch2010_UKoverview.pdf (accessed 8 April 2013).

Ekinsmyth, C. (2002) Project organisation, emdebedness and risk in magazine publishing, *Regional Studies*, 36(3), pp. 229–243.

Entwistle, J. & Wissinger, E. (2006) Keeping up appearances: Aesthetic labour in the fashion modelling industries of London and New York, *The Sociological Review*, 54(4), pp. 774–794.

European Commission. (2010) *Green Paper: Unlocking the Potential of the Creative Industries* (Brussels: European Commission).

European Commission. (2011) The entrepreneurial dimension of the cultural and creative industries. Available at http://ec.europa.eu/culture/key-documents/doc/studies/entrepreneurial/EDCCI_report.pdf (accessed 12 October 2011).

Faulkner, R. (1983) *Music on Demand: Composers and Careers in the Hollywood Film Industry* (NJ: Transaction Publishers).

Fieldhouse, R. (2007) GP freelancers—are chambers a locum team come true? *Management in Practice*, 11, pp. 32–33.

Glucker, J. (2006) A relational assessment of international market entry in management consulting, *Journal of Economic Geography*, 6(3), pp. 369–393.

Grabher, G. (2004a) Learning in projects, remembering in networks? Communality, sociality and connectivity in project ecologies, *European Urban and Regional Studies*, 11(2), pp. 103–123.

Grabher, G. (2004b) Temporary architectures of learning: Knowledge governance in project ecologies, *Organization Studies*, 25(9), pp. 1491–1514.

Gunnell, B. & Bright, M. (2010) *Creative Survival in Hard Times* (London: Arts Council).

Heery, E., Conley, H., Delbridge, R. & Stewart, P. (2004) Beyond the enterprise: Trade union representation of freelances in the UK, *Human Resource Management Journal*, 14(2), pp. 20–35.

Higgs, P., Cunningham, S. & Bakhshi, H. (2007) *Beyond the Creative Industries: Mapping the Creative Economy in the United Kingdom* (London: NESTA).

Jones, A. (2007) More than managing across borders? The complex role of face-to-face interaction in globalizing law firms, *Journal of Economic Geography*, 7(3), pp. 223–246.

Kamara, J., Augenbroe, G., Anumba, C. & Carrillo, P. (2002) Knowledge management in the architecture, engineering and construction industry, *Construction Innovation: Information, Process, Management*, 2(1), pp. 53–67.

Kitchings, J. & Smallbone, D. (2012) *Exploring the UK Freelance Workforce, 2011*. Available at http://www.pcg.org.uk/sites/default/files/media/documents/RESOURCES/01267%20PCG%20A4%2048PP%20KINGSTON%20REPORT%20WEB.PDF (accessed 8 April 2013).

Knell, J. & Oakley, K. (2007) *London's Creative Economy: An Accidental Success?* (London: Work Foundation).

Leyshon, A. (2011) Towards a non-economic, economic geography? From black boxes to the cultural circuit of capital in economic geographies of firms and managers, in: A. Leyshon, R. Lee, L. Mcdowell & P. Sunley (Eds) *The Sage Handbook of Economic Geography*, pp. 383–397 (London: Sage).

Malone, T. & Laubacher, R. (1998) Dawn of the e-lance economy, *Harvard Business Review*, 76 (Sept–Oct), pp. 145–152.

Mcrobbie, A. (2002) Clubs to companies: Notes on the decline of political culture in speeded up creative worlds, *Cultural Studies*, 16(4), pp. 516–531.

Mitchell, L. (2005) *Freelancing for Television and Radio* (London: Routledge).

Mould, O. (2008) Moving images: World cities, connections and projects in Sydney's TV production industry, *Global Networks*, 8(4), pp. 474–495.

NdotM (2011) *Make a Job, Don't take a Job*. Available at http://www.thecreativesociety.co.uk/wp-content/uploads/2011/02/the-creative-society-make-a-job-report.pdf (accessed 8 April 2013).

Nies, G. & Pedersini, R. (2003) Freelance journalists in the European media industry, *European Federation of Journalists*. Available at http://www.ifj.org/assets/docs/251/142/9d877fb-224c58e.pdf (accessed 8 April 2013).

O'Connor, J. (2007) The cultural and creative industries: A review of the literature. Available at www.artscouncil.org.uk (accessed 23 June 2008).

Pratt, A. (2008a) Cultural commodity chains, cultural clusters, or cultural production chains? *Growth and Change*, 39(1), pp. 95–103.

Pratt, A. (2008b) Creative cities: The cultural industries and the creative class, *Geografiska Annaler B*, 90(2), pp. 107–117.

Rifkin, J. (2005) When markets give way to networks, in: J. Hartley (Ed.) *The Creative Industries*, pp. 361–374 (Oxford: Blackwell).

Saundry, R., Stuart, M. & Antcliff, V. (2006) "It's more than who you know"—networks and trade unions in the audio-visual industries, *Human Resource Management Journal*, 16(4), pp. 376–392.

Skillset (2011) *Sector Skills Assessment for the Creative Media Industries in the UK*. Available at http://www.skillset.org/uploads/pdf/asset_16297.pdf?3 (accessed 8 April 2013).

Stanworth, C. & Stanworth, J. (1997) Managing an externalised workforce: Freelance labour-use in the UK book publishing industry, *Industrial Relations Journal*, 28(1), pp. 34–55.

Storey, J., Salaman, G. & Platman, K. (2005) Living with enterprise in an enterprise economy: Freelance and contract workers in the media, *Human Relations*, 58(8), pp. 1033–1054.

Taylor, M. & Asheim, B. (2001) The concept of the firm in economic geography, *Economic Geography*, 77(4), pp. 315–328.

Tempest, S. (2009) Learning from the alien: Knowledge relationships with temporary workers in network contexts, *The International Journal of Human Resource Management*, 20(4), pp. 912–927.

Vorley, T., Mould, O. & Lawton-Smith, H. (2008) Introduction to geographical economies of creativity, enterprise and the creative industries, *Geografiska Annaler B*, 90(2), pp. 101–106.

Weller, S. (2008) Beyond "global production networks": Australian fashion week's trans-sectoral synergies, *Growth and Change*, 39(1), pp. 104–122.

Yeung, H. (2001) Regulating 'the firm' and sociocultural practices in industrial geography II, *Progress in Human Geography*, 25(2), pp. 293–302.

Yeung, H. (2006) Firms, in: I. Douglas, R. Huggett & C. Perkins (Eds) *Companion Encyclopaedia of Geography: From Local to Global*, pp. 341–352 (London: Routledge).

The Role of Universities in the Regional Creative Economies of the UK: Hidden Protagonists and the Challenge of Knowledge Transfer

ROBERTA COMUNIAN*, CALVIN TAYLOR** & DAVID N. SMITH†

*Department of Culture, Media and Creative Industries, King's College London, London, UK, **School of Performance and Cultural Industries, University of Leeds, Leeds, UK, †The Centre for Research in Lifelong Learning, Glasgow Caledonian University, Glasgow, UK

ABSTRACT The Triple-Helix model of knowledge−industry−government relationships is one of the most comprehensive attempts to explain the changing institutional frameworks for innovation and growth, especially in the regional and urban contexts. Since the 1970s policies have been developed across Europe to evolve this institutional landscape. Since the late 1990s, regional and urban development strategies have also sought to harness the growth potential of the cultural and creative industries to regional and urban economic development. However, whilst the regional and urban planning literature has examined the growth-promoting potential of universities very closely, their possible role in relation to regional and urban creative economic development has received less attention. This paper aims to begin addressing this gap by interrogating the relationship between universities and the regional creative economy using, as a starting point, a model of analysis suggested by the Triple-Helix theoretical framework. The paper finds that whilst universities possess often long and hidden associations with regional and urban creative activities—as hidden protagonists—there are important institutional and professional challenges in the possibility of their developing an explicit and sustainable role as new actors in the regional and urban creative economies. The paper identifies the nature of these challenges with a view to developing a clearer understanding of the system, policy and institutional realities that underpin the often complex dynamics of knowledge creation−practice relationships found in arts and humanities disciplines.

1. Introduction

This paper uses the theoretical model of the Triple Helix (Etzkowitz & Leydesdorff, 1997; Leydesdorff & Etzkowitz, 1998; Etzkowitz & Leydesdorff, 2000; Leydesdorff &

Etzkowitz, 2001; Etzkowitz, 2003) to examine the potential relationship between two key phenomena within the regional and urban planning literatures: the role of the cultural and creative industries (CCIs) in fostering regional and urban innovation and growth, and the role of institutions of higher education in promoting these objectives. For the purposes of this paper, the CCIs are taken to be industrial activities that are primarily geared towards the production of symbolic products, the value of which is ultimately valorized in a market-place (Hesmondhalgh, 2007). The paper draws upon the experience of the UK, which, since the election of the Coalition Government in May 2010, has conspicuously withdrawn from the regional development agenda, and, in stark distinction with much of mainland Europe, has also disengaged from the CCI development agenda which has now become so much associated with previous New Labour governments.

Universities in the UK under both the previous and new regimes were and continue to be deeply embedded in knowledge economy policy discourse, both shaping it and being the recipients of specific funding to promote it (Charles, 2003; Harloe & Perry, 2004). The period 1997 through to 2008 also saw a high level of government policy activism (national, regional and local) on the regional and urban benefits of the CCIs, accompanied by a rich and highly varied research effort drawing on a wide range of disciplines and a wide range of research agents, including academics, policy analysts, consultants and CCI intermediary and lobby organizations (Hall, 2000; Jayne, 2005; Chapain & Comunian, 2010; Taylor, 2013). The starting point for the argument of this paper is the recognition that these two facets of public policy—regional policy-making on the one hand and the role of universities in the development of the creative industries on the other—have not yet explored their potential interaction and overlap.

The research literature on the role of universities in the innovation system is extensive and includes detailed studies on knowledge transfer and collaboration (Bercovitz & Feldman, 2006) and models of innovation and their key relationships (Dodgson *et al.*, 2005), but so far the research has concentrated on specific—mostly science and technology—disciplinary boundaries of university interaction with policy initiatives and the economy (Lindelöf & Löfsten, 2004). Similarly, a review of the research on the CCIs reveals many studies which analyse, for example, the cultural and creative production system (Pratt, 1997a, 1997b, 2004, 2008; Adkins *et al.*, 2007; Bakhshi & McVittie, 2009; Abadie *et al.*, 2010; Potts & Cunningham, 2010; Taylor, 2011), the role played by networks and informal relations (Banks *et al.*, 2000; Delmestri *et al.*, 2005; Adkins *et al.*, 2007; Antcliff *et al.*, 2007; Dahlstrom & Hermelin, 2007; Potts *et al.*, 2008; Currid & Williams, 2010; Lange, 2010; Lingo & O'Mahony, 2010), the importance of places and clusters (Bassett & Griffiths, 2002; Drake, 2003; Mommaas, 2004; Neff, 2005; Bathelt & Graf, 2008; Gwee, 2009; Pratt, 2009; Collis *et al.*, 2010; Thomas *et al.*, 2010) and the conditions and drivers of creative labour (Banks, 2006; Comunian, 2009). The role of HE institutions within this new CCI landscape in the UK has received some attention (Crossick, 2006; Powell, 2007; Taylor, 2007; Comunian & Faggian, 2011) but little with an explicitly regional and urban development focus.

The aims of this paper are two-fold: first, to begin creating a conceptual bridge between these two bodies of research which can inform future planning knowledge and understanding and, second, to contribute to the process of mapping possible models of interaction and the means by which CCI–university–government relationships might be promoted or inhibited. The advent of policies to promote interaction between businesses, institutions and public bodies has prompted the development of a range of models used to explain the changing regional and urban economic landscape. One of the most prominent of these, the Triple-Helix model, is used in the paper as an initial framework to begin the process of disclosing the emergent multiplex dynamics and interactions between the three spheres of the higher-education system, the CCIs and public policy. In particular, we are interested in what the model has to say about the dialectical relationship between recursive and reflexive modes of change and adaptation in the knowledge-innovation system. We stress that the framework is used here with the modest ambition of initiating possible avenues of analysis rather than, for example, the much more ambitious (and almost certainly contentious) project of proving any correspondence between the knowledge transfer dynamics of the CCIs and those of say the science and technology field. Since one of the central theoretical tenets of the Triple-Helix model is that of the generative and evolutionary power of relationships, this paper is particularly interested in how academics as a professional group central to the operation of the Triple Helix have responded to the seemingly intensifying interactions between the spheres of higher education, the CCIs and public policy. In particular, the paper bases part of its findings on a series of interviews undertaken with academics during the course of 2007 and 2008 which focused on the self-perceived and self-reported roles that academics play in the creative economy; the value they attribute to their interactions with creative businesses, organizations and practitioners, and what they see as the potential enablers for and barriers against such activities. A more detailed discussion on the methodology and data collected is presented in paragraph 3.1.

The paper is structured in three parts. The first sets out a brief synopsis of the key relevant elements of the Triple-Helix theoretical framework, drawing on central contributions in its formation. The second sets out how it might be used articulate the industry−policy−knowledge relationships of the creative economy, drawing on a range of research contributions on the CCIs. The third part of the paper presents and discusses the findings from empirical research undertaken with UK-based academics as key agents in the Triple Helix, framed according to the analysis presented in part two. Our findings suggest that universities have long interacted with their regional creative economies and, at least until very recently, have continued to expand their engagement. However, rather than the dialectically recursive and reflexive institutional adaptation advanced by the Triple-Helix model, what we find is that academic engagement with the creative economy is heavily mediated by three sets of qualifying phenomena: the structural expectations of the higher education system (Benner & Sandström, 2000; Lawton Smith, 2007), persistent institutional realities (of historic mission, academic organization and academic culture) and by the norms and values of discipline and academic professional practice (Bullen *et al.*, 2004). The paper principally aims to stimulate further debate by arguing for the need for a better understanding of the complex, sometimes explicit, often implicit, roles that institutions of higher education play in shaping their regional and urban creative economies. The Triple-Helix model is helpful in some respects as an important contribution to this objective, but, as we suggest later, it may be limited in some key areas.

2. The Triple Helix

The Triple-Helix model of industry−policy−knowledge relationships was introduced into the academic and policy worlds by the work of Leydesdorff and Etzkowitz (1996), who argued that these rich triplicate relationships were conspicuously influential in the shaping of systems for innovation and growth. Arguing against the familiar and much critiqued linear interpretation of knowledge creation, they explain that "a spiral model of innovation is required to capture multiple reciprocal linkages at different stages of the capitalization of knowledge" (Etzkowitz & Leydesdorff, 1997, p. 1). Observing that the present historical epoch is notable for its state of social, economic and cultural flux, innovation systems are increasingly structured, not by the prevailing institutional arrangements for innovation but by the interactions between agents and the systems of communication and intermediation (including new temporary organizations) they create to enable new innovation to take place (Etzkowitz & Leydesdorff, 2000). The sense of intense reflexivity this introduces into the system has the effect of de-centring traditional institutional arrangements, de-coupling institutions from their traditional functions and setting in motion an evolutionary process of functional combination and re-combination. In a very real sense, historic institutional certainties weaken, new narratives of purpose and intention are created, and new temporary communities of practice—and their necessary organizational arrangements—emerge and submerge according to the dialectic of recursion and reflexivity between the helices of the Triple Helix.

From the first seminal papers in 1996 and 1997, a dynamic research field has emerged (Fritsch & Schwirten, 1999; Lindelöf & Löfsten, 2004) expanding both its geographical reach and the range of sub-topics covered by Triple-Helix analysis. At the heart of the model is the key proposition that innovation springs from the "generative relationships" (Etzkowitz & Leydesdorff, 1997) created between agents and the transformations that ensue for both actors within the relationships and in the relationships themselves. As the commentators Viale and Pozzali (2010) observe, the value of the Triple Helix lies in the relationship between feedback and change. With this central tenet in mind, here are the four key features of the Triple-Helix model that we use in relation to our analysis of both the existing selected research and which inform our analysis of the interviews with representatives of the academic community.

2.1 *Mutliplex Relationships*

The first concerns that Etzkowitz and colleagues see as the proliferation of mutiplex relationships between the three spheres of knowledge, industry and government operative at differentiated scales, geographically, sectorally and politically (Etzkowitz and Leydesdorff, 2000). In their example, governments that were hitherto constrained to interact at the national level with industries under their own jurisdiction can now interact with sectors across a scale from international to local and vice versa.

2.2 *Evaluation*

The second concerns the model of outcome evaluation by which Triple-Helix interactions and actions are evaluated. The increased contingency and chance of the knowledge economy renders *ex ante* evaluation pretty much impossible. This places an increased stress on the need for quantifiable measures of *ex post* evaluation. Agents may not know the value of a particular interaction at its inception, and may indeed be prepared to entertain a wide variety of possible courses of action, but they do need to be in a position to evaluate it afterwards. As a result, evaluation tends inevitably towards the quantitative (Leydesdorff and Etzkowitz, 2000). This closely ties with the third characteristic.

2.3 *Organizational Innovation*

Increased contingency prompts institutions to develop more intuitive and improvizational strategies. These can take institutions outside their institutional comfort zones as helical combination and re-combination engenders a dialectical spiral of recursive institutional differentiation and reflexive institutional de-differentiation (Etzkowitz, 2003). This also applies to how the university can become a component element within the new spaces of innovation that have proliferated beyond the laboratory to encompass a wider range of metaphors for knowledge production and applications activities. The innovation landscape takes on new shapes—niches, clusters, *filieres*, *milieux*, etc. (Etzkowitz, 2003).

2.4 *Knowledge Exchange*

Historically, this revolves around "knowledge-push" and "market-pull", but in a fourth characteristic, the Triple-Helix model argues that distinctions such as "basic-applied" and "Mode I-Mode II" (Gibbons, 1994) may not be as absolute as their originators assume. Within the Triple Helix, multi-form possibilities are present. As Etzkowitz and Leydesdorff (2000) explain, the poles of these binaries are as likely to exist within each other as much as they are likely to co-exist in tension with each other.

With these four characteristics acting as lenses, we now turn to how this analysis of the Triple Helix might be used to examine the creative economy.

3. The Triple Helix and the Creative Economy

All of these characteristics are relevant to the study of regional and urban creative economies and their interactions with public policy and higher education. Figure 1 offers a

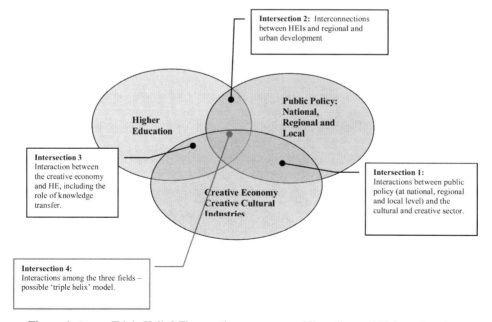

Figure 1. A new Triple Helix? The creative economy, public policy and higher education.

simple, provisional mapping schema of the Triple-Helix intersections (labelled 1–4) in regional and urban creative economies. This schema frames the reading of a range of research contributions that have already potentially paved the way for the development of a knowledge base supporting this model. This forms the core of the conceptual bridge that we think will enable these literatures to become connected. In particular, the following sections summarize: first, key research contributions on the relationship between government and local and regional public policies and the role attributed to the creative and cultural sector within these (labelled Intersection 1); second, a sample of theoretical research on the claimed role of higher education on the delivery of local and regional economic development and policies (labelled Intersection 2); third, a synoptic exploration of the conceptual potential of the Triple-Helix model of the relationships between higher education and the creative economy (labelled Intersection 3). After considering these first three intersections, we then provisionally consider the fourth intersection—the point where these three intersections overlap. This putative Triple-Helix model of the creative economy then becomes the focal concern of our examination of the empirical material in the third section of the paper.

3.1 *Intersection 1: The Creative and Cultural Industries and Public Policy*

The economic growth potential of the CCIs has animated the field of UK regional and urban development policy for at least 15 years (Bianchini & Parkinson, 1993; Griffiths, 1993; Bianchini & Landry, 1995; Griffiths, 1995a, 1995b; Pratt, 1997a, 1997b, 2004, 2005; Wynne & O'Connor, 1998; Brown *et al.*, 2000; Bassett & Griffiths, 2002; Griffiths *et al.*, 2003). This is now a major area of interest for European policy-makers (European Commission, 2005). However, this interest extends beyond the merely promotional. In an important reflexive move, the CCIs are as much a product of the constitutive power of the state's role in economic governance, as they are a product of secular industrial development. As O'Connor (1999) point out, the term "cultural industries" was first used extensively by the Greater London Council (GLC) in the 1980s as a rhetorical device designed to promote an "alternative economic" model of cultural policy.

In a move prompted as much by rhetoric (of a distinctively different hue) as by motives of indicative planning, the new UK central government of 1997 moved very quickly to recognize the "creative industries" by establishing the Creative Industries Task Force shortly after its election. The Task Force comprised representatives of the putative creative industries, including existing and nascent intermediary organizations and representatives of a number of government departments. One of its first actions, in a perfect example of institutional reflexivity set about mapping (describing and quantifying) the creative industries, and identifying policy measures that could promote their further development. The Creative Industries Mapping Document (DCMS, 1998a, 1998b) is one of the most quoted documents in the field. One of the main reasons for this is that alongside its seemingly prosaic descriptive and statistical concerns, it offered an analytical definition of the creative industries sector, famously described as "those activities which have their origin in individual creativity, skill and talent and which have the potential for wealth and job creation through the generation and exploitation of intellectual property" (DCMS, 1998a, 1998b, p. 3). Although the ensuing debate quickly descended into arguments over what should be included and what should be excluded, the important effect had occurred: government and industry (and some representatives of higher education and consultancy) had discursively constructed an object to which all parties could relate. Once established, the further constitution of the creative industries emerged in rapid piecemeal fashion in a series of reports commenting variously on television exports (DCMS, 1998a, 1998b), the contribution of the creative industries to national exports (DCMS, 1999a, 1999b), the internet (DCMS) and the regional dimension (DCMS, 1999a, 1999b), together with regular reviews of the economic contribution of the sector to the national economy.

Conspicuously, the constitutive power of public policy is revealed in another important dimension. In addition to the economic importance of the CCIs, a wide literature, both academic and public-policy-based, has explored how the sector impacts on developments in a wide range of fields, usually corresponding to the regional or local scale, including urban regeneration, social cohesion, civic participation, quality of life and revitalization (Bianchini & Landry, 1995; Griffiths, 1995a, 1995b; Markusen & Schrock, 2006). This milieu, however, evidences further Triple-Helix qualities. The boundaries between the types of organizations in the creative industries can often be blurred as commercial organizations take active roles in public interest activities and voluntary organizations take on more commercial functions. Boundaries can be blurred and conditions appear to interchange, a reflection in particular of the specific contexts of each component sub-system and the project-based nature of contractual relations found across much of the sector (Grabher, 2001; Neff *et al.*, 2005). According to Pratt (1997a, 1997b) understanding how the CCIs work from a demand-side perspective requires a focus on the role of networks and institutions and the social division of labour across firms. The force of social networks as market-forming has also been the subject of conceptual work on the creative industries (Potts *et al.*, 2008).

The policy implications that flow from these kinds of analyses are important for our argument. The CCIs draw together a wide network of agencies and stakeholders that range from the field of culture to the industrial and not-for-profit sectors, which together prompts speculation about the appropriate type of governance for these arrangements (Jeffcutt & Pratt, 2002). This shared governance and the role of networks across different sectors appears to suggest that the CCIs contain a high degree of connectivity both in the public infrastructure and in the production and consumption of economic outputs (Comunian, 2010). Jeffcutt and Pratt (2002) describe these arrangements as follows: "Hybrid and emergent organisational spaces, made up of dynamic interfaces between multiple stakeholders with many layers of knowledge are both characteristic of, and endemic in, the cultural industries" (Jeffcutt & Pratt, 2002, p. 231). Hybridity, emergence, multiple stakeholders and multiple layers of knowledge all point to qualities of the Triple Helix.

As boundaries shift between organization and network, public and private, market and voluntary sector, it is important to acknowledge how this multiplexity also became spatially inflected by the emphasis after 1997 in the UK on the regional and urban contexts. This regionalization agenda gave especial policy cachet to the CCIs. With this came a new twist in cultural policy as the CCIs were not only related to the then newly conceived economic role of regions, particularly in English political discourse, but to the economic arguments increasingly deployed to explain the creative potential and economic competitiveness of specific localities (Pratt, 2004), an understanding that also drove the focus of certain aspects of European Structural Fund intervention (Taylor, 2009). In the UK, the advocacy in support of the CCIs has been linked to a specifically regional development perspective leading to what a number of commentators have described as a fairly standard repertoire of policy constructs such as cultural quarters and creative clusters

(Jayne, 2005). The Sheffield Cultural Quarter, established in the early 1980s as a local government initiative, provides a pioneering example. As Moss (2002) suggests, the emphasis of the project was mainly on job creation in cultural production. However, other examples have been created in different cities around the UK such as Bristol (Bassett *et al.*, 2002), London (Newman & Smith, 2000) but also in smaller towns such as Huddersfield (Wood & Taylor, 2004) in West Yorkshire. In all these examples, local development agencies and local authority initiatives played central roles. Later developments continue the trend for public policy to have a defining role through its interaction with other agents. Indeed, the thrust of the later New Labour government's work on the creative industries reflected in the Creative Economy Programme (DCMS & BERR, 2008), together with a flurry of policy-related papers from various think-tanks suggests that the existence of the sector continues to result from the discursive effects of public policy (NESTA, 2007; The Work Foundation, 2008).

3.2 *Intersection 2: Higher Education and Regional Development*

There is extensive literature addressing the role of higher education in regional economic development and we will only comment on a small number of relevant topics here. Authors commonly recognize that this particular attention to the potential impact of higher education has been linked to a national knowledge economy agenda, an agenda to which the CCI policy agenda has an ambiguous relationship. Although it is difficult to summarize the complex role of institutions of higher education in a specific geographical context, the literature articulates three key dimensions for our purposes:

- *Human capital:* higher-education institutions contribute to a specific locality though the provision of graduates and a highly educated workforce (Florida, 1999). This human capital, although very mobile (Faggian & McCann, 2009), can influence the local economic development of specific contexts. Etzkowitz and Leydesdorff (2000) argue that the supply of graduates may in fact be universities' most important contribution to innovation;
- *Knowledge*: it is acknowledged that the knowledge generated by universities can through a variety of processes (knowledge transfer, spin-off companies, knowledge spillovers etc.) enrich the regional context (Audretsch *et al.*, 2005) and give raise to potential economic benefits derived by that knowledge (Anselin *et al.*, 2000). Universities can adopt more or less entrepreneurial approaches in managing these spillovers (Clark, 1998);
- *Infrastructure*: in the processes through which knowledge and human capital interact and contribute to the local context there is always an element of infrastructure development taking place. This might, for example, be a new incubator space (Rothaermel & Thursby, 2005) or new premises and conference facilities as well as new networking spaces or virtual platforms for interaction.

While much of the literature tends to concentrate on specific aspects of the impact of higher education and their interactions with the knowledge economy, many authors recognize the complexity of the knowledge interactions taking place. However, as Harloe and Perry (2004), for example, have argued the much-anticipated alignment of university interests with the knowledge economy agenda has at best been uneven, and possibly even unconvincing. They challenge the view that universities are moving seamlessly from "Mode 1" knowledge-production regimes (knowledge generated and controlled by specific disciplinary communities) to "Mode 2" regimes (where knowledge is generated and applied in a trans-disciplinary and applied way (Gibbons, 1994)). The picture, they suggest, appears much more complex with multiple and overlapping influences and interests at work. In many ways the engagement that universities have with the regional economy exhibit both traditional priorities and new inflexions of older educational agendas. Whilst in Intersection 1 we saw evidence of the Triple-Helix-like transformations, there is increased doubt about the extent to which universities as institutions are capable of creating the governance arrangements that would enable the full Triple-Helix model to work (Etzkowitz & Leydesdorff, 2000; Ozga & Jones, 2006).

3.3 *Intersection 3: The Role of Universities in the Regional Creative Economy*

It has been argued that universities have been long-term, but often "hidden" protagonists of the cultural economy, specifically at local and regional levels (Chatterton & Goddard, 2000). It is worth recalling that the majority of universities in the UK were in fact established to fulfil economic functions (Bond & Paterson, 2005) in ways which have often rendered the cultural functions of universities less visible. The foundation of the arts and humanities faculties was part of this process of cultural and creative engagement. Not only were the universities historically the training grounds for the professions, but by the 1950s industry was increasingly turning to arts graduates to solve the problem of a growing shortage of technologists (Sanderson, 1988). The contribution of the civic role of universities in developing the cultural life of many UK cities (Smith *et al.*, 2008) has been demonstrated in the commitment of a large part of the university workforce to cultural activities, their dissemination and, specifically in the areas of arts and humanities, the provision of expertise and infrastructure in what might historically be termed the "cultural third mission" (Smith, 2013).

In these roles, university museums and galleries have a long history of contributing to the local cultural offer, alongside the more contemporary university theatre and students union. If we read across from the regional development functions of universities to the CCIs, we quickly see that the *human capital* dimension has been the main focus of the recent literature, especially influenced by the work of Florida (Florida, 2002a, 2002b, 2002c, 2003). While in other disciplinary areas, universities are considered central to the regional economy because they engage actively in research exploitation through such activities as technology transfer, patenting and spin-offs, there is an inherent and not always welcome challenge for the arts and humanities research base (Bullen *et al.*, 2004). This is complicated by the very nature of the CCIs as an industrial sector—consisting of micro-businesses and with little capacity to finance or support external R&D—and which has implications for the knowledge and infra-structural roles ascribed in dominant innovation discourse to universities.

The CCIs sector comprises mainly small and medium-sized enterprises (SMEs) of less than 250 employees. 99.6% of UK CCI companies fall into this category, approximately 90% of the whole sector consists of companies with less than 10 employees and the few (around 500) large enterprises that do exist in the sector are concentrated mainly in London (Taylor, 2007). This raises two main questions. First, it implies the need to recognize regional differentiation within the creative economy, since universities located beyond the national capital city will almost certainly need to forge better relations with SMEs if they are to engage with the CCIs. Second, it raises questions about the potential affinity between particular kinds of higher-education institutions and the CCIs. Given the propensity of the research-intensive universities to define their missions internationally in terms of their research quality, striving for success in their local and regional relationships with the CCIs may require some structural and cultural plurality in approaches towards collaborations with the SME sector. As a result, even some of the most entrepreneurial universities have responded unevenly to the third mission of economic development and certainly from a system-wide point of view the picture is complex—and incomplete.

It has been suggested that the general challenges of institutional adaptation faced by universities are even more intense at the disciplinary level. Clark's (1998) study of entrepreneurial universities in different European countries argued that while science and technology departments had found it relatively easy (which may be more appearance than reality) to adjust to the new entrepreneurial regimes, arts and humanities departments could be characterized as the "resisting laggards" (Clark, 1998). Clark, incidentally, thought they might have good reason, since new money may not flow readily from government or non-government sources for these activities, reducing the incentive to change. Nevertheless, in a number of cases partial transformation has taken place resulting in some institutions existing in a "schizophrenic state, entrepreneurial on one side, traditional on the other" (Clark, 1998, p. 141). Developments over the last 10 years in UK research-funding models will undoubtedly provide future incentives for engagement.

Despite these apparent gaps, the evidence suggests actual wide-spread engagement. The national survey of English universities' interaction with business in 2001–2002 (HEFCE, 2003) found that one of the most commonly reported institutional intentions was to work with the CCIs. The subsequent high-profile Lambert Review of Business—University Collaboration (HM Treasury, 2003), commissioned by the UK Treasury noted:

> there are many excellent examples of collaborations involving the creative industries and universities or colleges of art and design. Policy-makers must ensure that policies aimed at promoting knowledge transfer are broad enough to allow initiatives such as these to grow and flourish, and that the focus is not entirely on science and engineering. (HM Treasury, 2003, p. 45)

Nevertheless, whilst a broad range of types of institution acknowledge their work with the CCIs, it was most marked within the "new" universities sector, a sector that also tends more explicitly to identify its purpose with the local and regional economies (HEFCE, 2003). The Higher Education Funding Council's own evaluation of its innovation and knowledge-transfer-funding programmes highlighted the unexpectedly high engagement in these activities by the arts.

Taking these three sets of summary observations into account, the synoptic picture at intersection 4 is complex; evidencing Triple-Helix-like processes within the CCIs, but more complexly arrayed structural and institutional priorities, motivations and expectations at the system and institutional levels of higher education. It is to this complexity that we now turn. The Triple Helix offers a sophisticated and nuanced account of how institutions may evolve, the reasons for that evolution and where it might lead. The world of UK higher education has seen intense policy action, specifically with regard to working with industry. In the next section we explore through interviews with 44 academics and academic leaders and managers how that new mission has been ingested into the institutional world of the university.

4. A New Triple Helix: Universities, Public Policy and the CCIs

This section interrogates the nature of a series of reported interactions between universities and the CCIs with special emphasis on those interactions that are with the arts and humanities research base. The methodology and data collected are presented before discussing the over-arching questions focused on: multiplexity, evaluation, organizational innovation, spatial innovation and knowledge exchange.

4.1 *Methodology, Data and Research Questions*

The 44 interviews were conducted during 2007 and 2008 with four sets of university personnel: executive leaders, departmental managers, central business development managers and academics. The sample of institutions selected was mapped against two criteria—regional location (ensuring a distribution by geography across the UK) and institutional mission. The acknowledged complexities of codifying the latter aside, the sample included specialist arts colleges, self-identified research-intensive universities and universities that fore-grounded their teaching and local industrial engagement missions. This produced a sample of 10 institutions (two pilot institutions and eight included in the principal sample). The interviews were conducted using a semi-structured interview schedule with four major areas of interest: the first covered the meanings attached by academic agents to the term "knowledge exchange" in the arts and humanities and the types of activities and engagements with external agencies included under such a rubric; the second covered the reported motivations offered by interviewees for undertaking these types of activities including perceptions of their value; the third area covered the ways in which such activities were supported, or where applicable, impeded; and, the fourth the ways in which value (academic and otherwise) is "captured" by institutions. Interviews were timed to last at least one hour with some variation. The majority of interviews were recorded by the interviewer and transcribed professionally. In three cases, hand-written notes were taken. Interviewees were selected from art, music and performance as represented in standard UK university subject classifications and produced the following structured sample (Table 1).

It is readily apparent from university promotional materials and sources, as well as from reports by sector bodies, that universities in the UK have taken on board the push for supporting the CCIs from public policy. Obvious signs of engagement are the development of specific CCI departments; new courses (especially at post-graduate level) aimed at creative entrepreneurship and innovation (Warwick, King's College London, Goldsmith's, and Leeds, for example) and growing centres of research with a specialist interest in the CCIs (at various times: Manchester Metropolitan, Goldsmith's College London, the London School of Economics, Leeds, King's College London and Warwick). Historical affiliations between programmes in music, fine arts, performance and design are also being re-tooled to reflect the broader significance of the CCIs, with the inclusion of enterprise education and work-based learning in the CCIs in the under-graduate curricula (Brown, 2007). This activity is mirrored in new funding programmes, especially for developing collaborations between university academics and CCI business and organizations,

Table 1. Summary of sample institutions

Institutional reference	Executive	Departmental	Central business unit	Total
Pilot				
A		3	1	4
B	1	1	2	4
Roll-out				
C	1	5		6
D		5		5
E	1	5	1	7
F	1	2		3
G	1	3		4
H	1	3		4
I	1	3		4
J	1	1	1	3
Total	8	31	5	44

including national initiatives such as the Creative Industries Knowledge Transfer Network, The AHRC Creative Economy Hubs and the regional London Centre for Arts and Cultural Enterprise (now The Culture Capital Exchange). Active UK funding bodies include the Technology Strategy Board, the Environmental and Physical Sciences Research Council, the Arts and Humanities Research Council, the Economic and Social Research Council and the Knowledge Transfer Partnership Scheme.

But how do institutions and academics describe their interactions with the CCI sector? What are the typical characteristics of such interactions? What encourages or inhibits them? How is value captured from them? From our interviews we have two broad conclusions, each with particular detail sensitized by the relative positions of the personnel interviewed and by ideas about institutional mission.

First is the conclusion that universities have unanimously taken on the "CCI proposition" in a wide range of ways, but especially highlighting the complex network of relations and interactions that characterize their knowledge exchange with creative practitioners and companies.

The second is that the nature of the interactions undertaken as a result and the ways in which they are valued are heavily mediated by what might be described as system and institutional realities (of funding, organizational structure, institutional culture and tradition) and disciplinary cultures (shared ideas across the academic—practice nexus about such matters as artistic credibility, professional repute, disciplinary values and norms). These mediating features persist as political and organizational realities in spite of attempts by policy-makers and funders to persuade universities and academics to adapt their behaviours and priorities in the direction of knowledge exploitation and application. Another key dimension that was of concern both for managers and academics was the issue of evaluation and how to capture the impact of what was taking place within and outside academia. Many of the respondents identified these knowledge practice and exchanges taking place as a new and evolving landscape requiring organizational innovation and mechanisms of learning and adaption. The respondents also provided interesting takes on the definition and practice of knowledge exchange, which critically engaged with the notions of knowledge transfer and Triple Helix.

4.2 *Multiplex Relationships*

Our interviews evidence a rapid expansion and diversification of the relationships that institutions, departments and individual academics have developed with organizations and individuals in the CCIs. However, they also show that any external push towards speeding up and reframing the collaboration and exchanges with industry and policy are set within the long-term practices of institutions, departments, disciplines and individuals. Three sets of considerations were regularly cited by academics in relation to interactions with the CCIs. These were: the place of external engagement within the professional academic identity, the relationship between the economy of academic esteem and practitioner reputation and the complex problematics of pursuing academic work in the arts and humanities which is sensitized to external engagement with departmental and institutional resource requirements.

Academics who work with other sectors often reported complexities and sometimes tensions in being able to fulfil an external mission which they see as being encouraged by funding streams and policy priorities, but which is still not seen by institutions as impacting upon notions of academic identity:

> I think that definition of the inside and the outside is the thing which has perhaps characterised what I've been doing all along. I did a talk ... which was attempting to deal with the edges of the institution, the inside and the outside and how one kind of managed that sort of interface, because it strikes me that that's one of the key problems in this area. And having been in a situation where I did sort of straddle that .... I felt acutely the difficulties which arose out of that. (Lecturer)

Responding to encouragement to engage in external activities was even reported as having limiting effects on careers: "Well I mean I think probably in career terms it didn't do me any good at all" (Lecturer).

4.3 *Evaluation*

The executive managers and leaders in universities we interviewed were amply aware of the public policy push behind the encouragement for universities to interact with the CCIs. That is readily evident in the willingness to validate new courses, create new departments and foster strategic alliances with leading CCI organizations. Interestingly, however, the leaders see these interactions as less as the university adapting to post-industrial economic and social forms and their associated priorities, but more as the university assimilating the CCIs into an on-going institutional narrative about locale and the civic role of the university, which forms a key component of the way that such interactions are evaluated. Often invoking narratives of origin, executive leaders see the CCIs as a new opportunity for the university to be seen as exercising its historic social obligations to locality. This is inflected in two ways depending on the nature of the institution. In some, this adoption is part of a historic narrative about the relationship between the university and local industry. Where universities have a strong sense of their connection with local industry—many were set up by groups of local entrepreneurs ("our connections with industry go back forever", Institutional Leader)—the approach to the CCIs is couched in terms of serving the local economy, particularly in terms of likely graduate destinations:

Where arts and humanities have come into commercialisation and consultancy has been in cross-over work really between what they do and what our Careers Advisory Service do around entrepreneurship education and start-ups for graduates.

In others, the narrative is couched in terms of the university as a historic patron of the arts and culture. This was particularly true in research-intensive institutions where cultural paternalism with respect to the arts sits alongside otherwise hard-edged knowledge economy narratives of intellectual property exploitation and industrial innovation. Specialist institutions, however, had a clear sense of both the contexts within which creative practice takes place, and of their own responsibilities to it, but also the inherent difficulties:

> It's challenging yeah. And they're very small scale businesses and so in knowledge transfer and buying services from us they're never going to be in a position to do it. (Executive Leader)

Departmental heads also share this broader strategic sense of the university in relation to the CCIs. However, their view is tempered by resource considerations and what was clearly a more institutionally pragmatic outlook:

> ... like every university, we are under pressure to bring in more income and we spent a lot of time last year developing new ideas for short courses and conferences and we may have some very imaginative ideas ... but in the end we just couldn't live with any of these plans because we couldn't make them sufficiently price competitive. (Departmental Head)

Typically as the authority accountable for resources, departmental heads find themselves in negotiation between the strategic imperatives of the institution, especially with respect to income generation and the achievement of core goals in relation to learning and teaching and research. However, for experienced heads of department there is also something resonant about the emergence of the "knowledge transfer agenda":

> It's of course been happening endlessly. So, what used to be coming back to the university to talk to a member of staff who used to be in a theatre company and now is setting one up and we're developing and, etc, etc. that's now formal mentoring with somebody. (Departmental Head)

This interaction between professional academic work and working in the creative industries has been seen by some departmental heads as offering the prospect of being able to capture benefit for the department, not always with the success sought:

> ... we have another member of staff who is only half time for us and who came to us with massive experience in cultural programming for the BBC ... and we always hoped and expected that we'd get a slice of that action, but we didn't, he's always been able to keep them very separate. (Departmental Leader)

4.4 *Organizational Innovation*

Such experiences as that just illustrated point to an organizational challenge. Engagements with the CCIs are open-ended, managed within disciplinary and local networks with academic and practice-based memberships, and where engagement does not always have tangible benefit to the university. Such membership networks provide both academics and practitioners with space to meet, share ideas and re-enforce disciplinary norms and values, especially in relation to matters of esteem, reputation and credibility. However, benefits back to the department are typically uncertain, unpredictable and generally unquantifiable in terms of likely payback. The opportunities for student engagement were unanimously supported with respondents very clearly endorsing the widespread adoption across the arts of agendas on employability, enterprise education and work-based learning. Engagements with external organizations prepared to provide placements, internships and other forms of work-based learning were particularly valued.

What this illustrates is that although universities understand that the CCIs are a deeply networked sector, they find it difficult to take the next Triple-Helix step and take on the "role of the other" by taking on the priorities, values and ways of working of the CCIs. Other solutions are improvised. As one music academic explains:

> Liaison between the culture industry or the popular music industry or the commercial industry and academia is fraught ... there aren't meeting places .... And so there are ... word of mouth ... exchanges that take place between commercial music and the public subsidised bodies and we're all interested in each other. We can talk to each other ... and I look for opportunities. I'm comfortable with that. Brokerage might be the key. (Lecturer)

However, there are structural inhibitors that highlight the set of asymmetries between university departments and CCI organizations:

> ... we like the idea of emulating science and technology, we like the idea of being organised and setting up large umbrella schemes to work within and I think that is possible. But actually the size of the companies are different ... it's quite difficult to find a creative company of the size that would support the sort of large-scale projects that are going on in science and technology. (Departmental Head)

4.5 *Knowledge Exchange*

The concept of knowledge exchange has taken over from the older concept of knowledge transfer in order to reflect the reflexive nature of university − industry relationships. It aligns closely with the dynamic of the Triple Helix. Academics were especially sensitized to both the discursive power of this shift in language—but also to the practical implications of what it might entail. The relationship between intellectual and artistic and cultural practices meshed complexly with imperatives for external engagement. As a music academic explained

> I think that the stuff, the matter, the material of our subject is, in itself, a form of knowledge transfer in any case because we're thinking about the world of ideas

... we're living in a world of communication and a world of critiquing one form and reading art forms and all this is related to the transference of knowledge ... We've always been engaged with the relationship between what we are researching and the audiences that are receiving it. (Lecturer)

In some instances the very nature of the activity contains within it both intellectual and practical knowledge, basic and applied knowledge and also the sense of an exchange relationship between the producers and consumers of cultural experience. In this sense we can see that aspect of the Triple Helix in which different modes of knowledge production sit alongside (and even within) each other.

Institutions were also highly sensitized to how the concept of knowledge exchange might be seen as combining a range of agendas with, interestingly, a key role for students. As one institutional leader explained:

Well I think it's used at the moment in probably three or so different ways and I don't think they're mutually exclusive. One obviously is a question of third stream and income generation and that's the most difficult to address. The other is around, increasingly now, around the idea of the other end of the spectrum if you want to call it, is to do with student enterprise and sort of developing notions of what we mean by student enterprise what with the Leitch Report and questions to do with employability. (Executive Leader)

5. Conclusion

The paper has tried to test the value of engaging with the existing models and literature on knowledge engagement and platforms for its support across academia, industry and public policy in relation to arts and humanities research. In particular, it has explored the practices of academics in arts and humanities through the lens of the Triple-Helix model, to further unfold a complex network of stakeholders and dynamics that have so far received very little attention in the literature.

The Triple-Helix framework has enabled us to highlight the role played by public policy in the creative economy and the specific nature—often small and fragmented—of the creative industries themselves to better understand the long-established but often informal interconnections with higher education in general and specially arts and humanities. While the informality and unstructured nature of most of the relationships described in the paper might seem to contradict some of the formal structure and dynamics described by Etzkowitz and Leydesdorff in their research on the Triple Helix, these appear only as superficial incongruities.

As Etzkowitz and Leydesdorff (2000) suggest, the Triple Helix describes not only the relationship between university, industry and government, but also the forms of internal transformations that can occur within these different spheres. In this respect it can be clearly argued that the model can be seen to be at work in the CCI sector and has a role in promoting a better understanding of how arts and humanities-based disciplines are engaging in knowledge transfer and exploitation activities with the wider CCIs sector. It can help researchers and academics to appreciate the dynamics of these relationships alongside those they create for teaching and researching. These relationships point to knowledge sharing, economic impact, knowledge spill-overs and local economic development, although much of it happens through experimentation, fluid structures and interconnection rather than formal platforms for interactions.

From our interviews we received two broad conclusions, each with particular detail sensitized to the relative institutional positions of the personnel interviewed and by ideas about institutional mission. The first is that diversity characterizes the ways that universities have taken on the "CCI proposition" as expressed through various forms of engagement or knowledge exchange between the research base and "knowledge" users: this is a complex knowledge creation − practice dynamic. The second is that the nature of the interactions undertaken as a result and the ways in which they are valued is heavily mediated by what might be described as system and institutional realities (of funding, organizational structure, institutional culture and tradition) and disciplinary cultures (shared ideas across the academic − practice nexus about such matters as artistic credibility, professional repute, disciplinary values and norms). These mediating features persist as political and organizational realities in spite of attempts by policy-makers and funders to persuade universities and academics to adapt their behaviours and priorities in the direction of knowledge exploitation and application.

However, on the basis of our interview work at 10 UK universities there are a number of considerations that universities and wider public policy need to make if they are to engage with this sector in a productive way. They have to consider whether and how their own processes of knowledge production and dissemination are appropriate for the creative industries. In a sector where tacit knowledge plays such a crucial role, this and other forms of untraded interdependency in regional development point to the need for universities to examine their own role in the creation and circulation of knowledge. Equally, whilst public policy appears to be pro-active in the promotion of regional development strategies incorporating the CCIs, what it might also need to consider more specifically are the characteristics of innovation in this sector and in particular its social dynamics, especially in relation to the role of tacit knowledge. The CCIs in particular trade heavily on the role of social interaction and, as Nevarez (2003) claims, universities may make more appropriate "chambers of commerce" for the creative industries than those of the traditional variety.

The key question for the arts and humanities disciplines concerns their relationship with new paradigms of knowledge production. Far from being an ill-fitting exceptional case in the knowledge economy, it may be that the interactions between the arts and humanities research base in higher education and the CCIs is actually defining and giving meaning to new knowledge exchange processes through new forms of organization, partnership, transdisciplinarity, accountability and reflexivity—new contexts of knowledge creation and diffusion. The heuristics of the Triple Helix provide a valuable opportunity to map this landscape and to open a new dialogue about the nature of knowledge production, transfer and exploitation in a sector that is in the process of rapid transformation. However, what this may signal is a growing expansion of the function of university research rather than necessarily a re-orientation of its purpose. To that end we see the moves being made by universities to shed their historic hidden protagonist guise and take on the mantle of active regional agents, not as a re-functionalization but as an assimilation of ostensibly new agendas to historic regimes of value derived from the academy as a particular kind of institution.

This paper has opened the way for more of this debate to take place; we feel that there are a number of interesting venues still to research. First, while recent publications

(Universities UK, 2010; AHRC, 2011) have explored the attitude and practice of arts and humanities academics towards external engagement, there is still very limited knowledge about the user-led engagement and the attitude of creative industries towards academia. Second, public and governmental organizations have been essential in providing support towards the development of creative industries; however, they have been slower in bridging academic research to the creative economy and creating collaborative frameworks. A better understanding of the role played by policy in supporting these creative connections is needed. Finally, the role of the engaged academic and specifically in the creative fields of the teaching practitioner is a key characteristic of the arts and humanities and needs to be better understood.

References

Abadie, F., Friedewald, M. & Weber, M. (2010) Adaptive foresight in the creative content industries: Anticipating value chain transformations and need for policy action, *Science and Public Policy*, 37(1), pp. 19–30.

Adkins, B., Foth, M., Summerville, J. & Higgs, P. L. (2007) Ecologies of innovation—symbolic aspects of cross-organizational linkages in the design sector in an Australian inner-city area, *American Behavioral Scientist*, 50(7), pp. 922–934.

AHRC (2011) *Hidden Connections—Knowledge Exchange Between Arts and Humanities and the Private, Public and Third Sectors* (Swindon: Arts and Humanities Research Council).

Anselin, L., Varga, A. & Acs, Z. (2000) Geographical spillovers and university research: A spatial econometric perspective, *Growth and Change*, 31(4), pp. 501–515.

Antcliff, V., Saundry, R. & Stuart, M. (2007) Networks and social capital in the UK television industry: The weakness of weak ties, *Human Relations*, 60(2), pp. 371–393.

Audretsch, D. B., Lehmann, E. E. & Warning, S. (2005) University spillovers and new firm location, *Research Policy*, 34(7), pp. 1113–1122.

Bakhshi, H. & McVittie, E. (2009) Creative supply-chain linkages and innovation: Do the creative industries stimulate business innovation in the wider economy? *Innovation-Management Policy & Practice*, 11(2), pp. 169–189.

Banks, M. (2006) Moral economy and cultural work, *Sociology*, 40(3), pp. 455–472.

Banks, M., Lovatt, A., O'Connor, J. & Raffo, C. (2000) Risk and trust in the cultural industries, *Geoforum*, 31(4), pp. 453–464.

Bassett, K. & Griffiths, R. (2002) Cultural industries, cultural clusters and the city: The example of natural history film-making in Bristol, *Geoforum*, 33, pp. 165–177.

Bassett, K., Griffiths, R. & Smith, I. (2002) Cultural industries, cultural clusters and the city: The example of natural history film-making in Bristol, *Geoforum*, 33(2), pp. 165–177.

Bathelt, H. & Graf, A. (2008) Internal and external dynamics of the Munich film and TV industry cluster, and limitations to future growth, *Environment and Planning A*, 40(8), pp. 1944–1965.

Benner, M. & Sandström, U. (2000) Institutionalizing the Triple Helix: Research funding and norms in the academic system, *Research Policy*, 29(2), pp. 291–301.

Bercovitz, J. & Feldman, M. (2006) Entreprenerial universities and technology transfer: A conceptual framework for understanding knowledge-based economic development, *The Journal of Technology Transfer*, 31(1), pp. 175–188.

Bianchini, F. & Landry, C. (1995) *The Creative City* (London: Demos).

Bianchini, F. & Parkinson, M. (Eds) (1993) *Cultural Policy and Urban Regeneration: The West European Experience* (Manchester: Manchester University Press).

Bond, R. & Paterson, L. (2005) Coming down from the ivory tower? Academics' civic and econonomic engagement with the community, *Oxford Review of Education*, 31(3), pp. 331–351.

Brown, R. (2007) Promoting entrepreneurship in arts education, in: C. Henry (Ed.) *Entrepreneurship in the Creative Industries: An International Perspective*, pp. 126–141 (Cheltenham: Edward Elgar).

Brown, A., O'Connor, J. & Cohen, S. (2000) Local music policies within a global music industry: Cultural quarters in Manchester and Sheffield, *Geoforum*, 31(4), pp. 437–451.

Bullen, E., Robb, S. & Kenway, J. (2004) Creative destruction: Knowledge economy policy and the future of the arts and humanities in the academy, *Journal of Education Policy*, 19(1), pp. 3–22.

Chapain, C. A. & Comunian, R. (2010) Enabling and inhibiting the creative economy: The role of the local and regional dimensions in England, *Regional Studies*, 43(6), pp. 717–734.

Charles, D. (2003) Universities and territorial development: Reshaping the regional role of English universities, *Local Economy*, 18(1), pp. 7–20.

Chatterton, P. & Goddard, J. (2000) The response of higher education institutions to regional needs, *European Journal of Education*, 35(4), pp. 475–496.

Clark, B. R. (1998) The entrepreneurial university: Demand and response, *Tertiary Education and Management*, 4(1), pp. 5–16.

Collis, C., Felton, E. & Graham, P. (2010) Beyond the inner city: Real and imagined places in creative place policy and practice, *Information Society*, 26(2), pp. 104–112.

Comunian, R. (2009) Questioning creative work as driver of economic development: The case of Newcastle-Gateshead, *Creative Industries Journal*, 2(1), pp. 57–71.

Comunian, R. (2010) Rethinking the creative city: The role of complexity, networks and interactions in the urban creative economy, *Urban Studies*, 48(6), pp. 1157–1179.

Comunian, R. & Faggian, A. (2011) Higher education and the creative city, in: C. Mellander, A. Andersson & D. Andersson (Eds) *Handbook on Cities and Creativity*, pp. 187–210 (London: Edward Elgar).

Crossick, G. (2006) *Knowledge Transfer Without Widgets: The Challenge of the Creative Economy. Lecture, Royal Society of Arts* (London: Goldsmiths University of London).

Currid, E. & Williams, S. (2010) Two cities, five industries: Similarities and differences within and between cultural industries in New York and Los Angeles, *Journal of Planning Education and Research*, 29(3), pp. 322–335.

Dahlstrom, M. & Hermelin, B. (2007) Creative industries, spatiality and flexibility: The example of film production, *Norsk Geografisk Tidsskrift-Norwegian Journal of Geography*, 61(3), pp. 111–121.

DCMS (1998a) *Creative Industries Mapping Document* (London: Department for Culture, Media and Sport).

DCMS (1998b) *The Report of the Creative Industries Taskforce into Television Exports* (London: Department for Culture, Media and Sport).

DCMS (1999a) *Creative Industries—The Regional Dimension* (London: Department for Culture, Media and Sport).

DCMS (1999b) *Creative Industries Exports—Our Hidden Potential* (London: Department for Culture, Media and Sport).

DCMS & BERR (2008) *Creative Britain—New Talents for the Economy* (London: DCMS).

Delmestri, G., Montanari, F. & Usai, A. (2005) Reputation and strength of ties in predicting commercial success and artistic merit of independents in the Italian feature film industry, *Journal of Management Studies*, 42(5), pp. 975–1002.

Dodgson, M., Gann, D. & Salter, A. (2005) *Think, Play, Do: Technology Innovation and Organisation* (Oxford: Oxford University Press).

Drake, G. (2003) This place gives me space: Place and creativity in the creative industries, *Geoforum*, 34(4), pp. 511–524.

Etzkowitz, H. (2003) Innovation in innovation: The triple helix of university-industry-government relations, *Social Science Information*, 42(3), pp. 293–337.

Etzkowitz, H. & Leydesdorff, L. (Eds) (1997) *Universities and the Global Knowledge Economy: A Triple Helix of University-Industry-Government Relations* (London: Pinter).

Etzkowitz, H. & Leydesdorff, L. (2000) The dynamics of innovation: From national systems and "mode 2" to a Triple Helix of university-industry-government relations, *Research Policy*, 29(2), pp. 109–123.

European Commission (2005) *The Future of the Creative Industries* (Brussels: European Commission).

Faggian, A. & McCann, P. (2009) Universities, agglomerations and graduate human capital mobility, *Journal of Economic and Social Geography (TESG)*, 100(2), pp. 210–223.

Florida, R. (1999, Summer) The role of the university: Leveraging talent, not technology, *Science and Technology*, 15(4), pp. 67–73.

Florida, R. (2002a) Bohemia and economic geography, *Journal of Economic Geography*, 2(1), pp. 55–71.

Florida, R. (2002b) The economic geography of talent, *Annals of the Association of American Geographers*, 92(4), pp. 743–755.

Florida, R. (2002c) *The Rise of the Creative Class* (New York: Basic Books).

Florida, R. (2003) Cities and creative class, *City and Community*, 2(1), pp. 3–19.

Fritsch, M. & Schwirten, C. (1999) Enterprise-university co-operation and the role of public research institutions in regional innovation systems, *Industry & Innovation*, 6(1), pp. 69–83.

Gibbons, M. L., Limoges, C., Nowotny, S., Schwartzmann, S., Scott, P. & Trow, M. (Eds) (1994) *The New Production of Knowledge: The Dynamics of Science and Research in Contemporary Societies* (London: Sage).

Grabher, G. (2001) Ecologies of creativity: The village, the group, and the heterarchic organisation of the British advertising industry, *Environment & Planning A*, 33(2), pp. 351–374.

Griffiths, R. (1993) The politics of cultural policy in urban regeneration strategies, *Policy and Politics*, 21(1), pp. 39–46.

Griffiths, R. (1995a) Cultural strategies and new modes of urban intervention, *Cities*, 12(4), pp. 253–265.

Griffiths, R. (1995b) The politics of cultural policy in urban regeneration strategies, *Policy and Politics*, 21(1), pp. 39–46.

Griffiths, R., Bassett, K. & Smith, I. (2003) Capitalising on culture: Cities and the changing landscape of cultural policy, *Policy and Politics*, 31(2), pp. 153–169.

Gwee, J. (2009) Innovation and the creative industries cluster: A case study of Singapore's creative industries, *Innovation-Management Policy & Practice*, 11(2), pp. 240–252.

Hall, P. (2000) Creative cities and economic development, *Urban Studies*, 37(4), pp. 639–649.

Harloe, M. & Perry, B. (2004) Universities, localities and regional development: The emergence of the "mode 2" university? *International Journal of Urban and Regional Research*, 28(1), pp. 212–223.

HEFCE (2003) *2001–2002 Higher Education Business and Community Interaction Survey* (London: Higher Education Funding Council for England (HEFCE)).

Hesmondhalgh, D. (2007) *The Cultural Industries* (London: Sage).

HM Tr & easury (2003) *The Lambert Review of Business-University Collaboration* (London: The Stationary Office).

Jayne, M. (2005) Creative industries: The regional dimension? *Environment & Planning C: Government & Policy*, 23(4), pp. 537–556.

Jeffcutt, P. & Pratt, A. C. (2002) Managing creativity in the cultural industries, *Creativity and Innovation Management*, 11(4), pp. 225–233.

Lange, B. (2010) Beyond creative production networks. The development of intra-metropolitan creative industries clusters in Berlin and New York City, *Zeitschrift Fur Wirtschaftsgeographie*, 54(2), pp. 140–142.

Lawton Smith, H. (2007) Universities, innovation, and territorial development: A review of the evidence, *Environment and Planning C: Government and Policy*, 25(1), pp. 98–114.

Leydesdorff, L. & Etzkowitz, H. (1996) Emergence of a triple helix of university-industry—government relations, *Science and Publich Policy*, 23, pp. 279–286.

Leydesdorff, L. & Etzkowitz, H. (1998) The Triple Helix as a model for innovation studies, *Science and Publich Policy*, 25(3), pp. 195–203.

Leydesdorff, L. & Etzkowitz, H. (2001) The transformation of university-industry-government relations. Available at http://www.sociology.org/content/vol005.004/th.html (accessed 5 June 2006).

Lindelöf, P. & Löfsten, H. (2004) Proximity as a resource base for competitive advantage: University-industry links for technology transfer, *Journal of Technology Transfer*, 29(3–4), pp. 311–326.

Lingo, E. L. & O'Mahony, S. (2010) Nexus work: Brokerage on creative projects, *Administrative Science Quarterly*, 55(1), pp. 47–81.

Markusen, A. & Schrock, G. (2006) The artistic dividend: Urban artistic specialisation and economic development implications, *Urban Studies*, 43(10), pp. 1661–1686.

Mommaas, H. (2004) Cultural clusters and the post-industrial city: Towards the remapping of urban cultural policy, *Urban Studies*, 41(3), pp. 507–532.

Moss, L. (2002) Sheffield's cultural industries quarter 20 years on: What can be learned from a pioneering example? *International Journal of Cultural Policy*, 8(2), pp. 211–219.

Neff, G. (2005) The changing place of cultural production: The location of social networks in the digital media industry, *Annals of the American Academy of Political and Social Science*, 597(1), pp. 134–152.

Neff, G., Wissinger, E. & Zukin, S. (2005) Entrepreneurial labor among cultural producers. "Cool" jobs in "Hot" industries, *Social Semiotics*, 15(3), pp. 307–334.

NESTA (2007) *How Linked are the UK's Creative Industries to the Wider Economy? An Input-Output Analysis*, W. Paper (London: NESTA).

Nevarez, L. (2003) *New Money, Nice Town: How Capital Works in the New Urban Economy* (London: Routledge).

Newman, P. & Smith, I. (2000) Cultural production, place and politics on the South Bank of the Thames, *International Journal of Urban and Regional Research*, 24(1), pp. 9–24.

O'Connor, J. (1999) *Definition of Cultural Industries* (Manchester: Manchester Metropolitan University, Manchester Institute for Popular Culture).

Ozga, J. & Jones, R. (2006) Travelling and embedded policy: The case of knowledge transfer, *Journal of Education Policy*, 21(1), pp. 1–17.

Potts, J. & Cunningham, S. (2010) Four models of the creative industries, *Revue D Economie Politique*, 120(1), pp. 163–180.

Potts, J., Cunningham, S., Hartley, J. & Ormerod, P. (2008) Social network markets: A new definition of the creative industries, *Journal of Cultural Economics*, 32(3), pp. 167–185.

Powell, J. (2007) Creative universities and their creative city-regions, *Industry and Higher Education*, 21(6), pp. 323–335.

Pratt, A. C. (1997a) The cultural industries production system: A case study of employment change in Britain, 1984-91, *Environment and Planning A*, 29(11), pp. 1953–1974.

Pratt, A. C. (1997b) Production values: From cultural industries to the governance of culture, *Environment and Planning A*, 29(11), pp. 1911–1917.

Pratt, A. C. (2004) Creative clusters: Towards the governance of the creative industries production system? *Media International Australia*, (112), pp. 50–66.

Pratt, A. C. (2005) City of quarters: Urban villages in the contemporary city, *Journal of Rural Studies*, 21(4), pp. 492–493.

Pratt, A. C. (2008) Creative cities: The cultural industries and the creative class, *Geografiska Annaler Series B-Human Geography*, 90B(2), pp. 107–117.

Pratt, A. C. (2009) Urban regeneration: From the arts "feel good" factor to the cultural economy: A case study of Hoxton, London, *Urban Studies*, 46(5–6), pp. 1041–1061.

Rothaermel, F. T. & Thursby, M. (2005) University-incubator firm knowledge flows: Assessing their impact on incubator firm performance, *Research Policy*, 34(3), pp. 305–320.

Sanderson, M. (1988) Education and economic decline, 1890–1980s, *Oxford Review of Economic Policy*, 4(1), pp. 38–50.

Smith, D. N. (2011) Academics, the "cultural third mission" and the BBC: Forgotten histories of knowledge creation, transformation and impact, *Studies in Higher Education*. doi: 10.1080/03075079.2011.594502.

Smith, D., Taylor, C. & Comunian R. (2008) Universities in the cultural economy: Bridging innovation in arts and humanities and the creative industries ICCPR 2008. International Conference on Cultural Policy Research, Istanbul, Turkey.

Taylor, C. (2007) Developing relationships between higher education, enterprise and innovation in the creative industries, in: C. Henry (Ed.) *Entrepreneurship in the Creative Industries—An International Perspective* (Cheltenham: Edward Elgar).

Taylor, C. (2009) The creative industries, governance and economic development: UK perspectives, in: L. Kong & J. O'Connor (Eds) *Creative Economies, Creative Cities: Asian-European Perspectives*, pp. 153–166 (Dordrecht: Springer).

Taylor, C. (2011) Socialising creativity: Entrepreneurship and innovation in the creative industries, in: C. Henry & A. de Bruin (Eds) *Entrepreneurship and the Creative Economy: Process, Practice and Policy*, pp. 30–49 (Cheltenham: Edward Elgar).

Taylor, C. (2013) Between culture, policy and industry: The modalities of intermediation in the creative economy, *Regional Studies*. doi: 10.1080/00343404.2012.748980.

The Work Foundation (2008) *Staying Ahead: The Economic Performance of the UK's Creative Industries* (London: The Work Foundation).

Thomas, N. J., Hawkins, H. & Harvey, D. C. (2010) The geographies of the creative industries: Scale, clusters and connectivity, *Geography*, 95(1), pp. 14–21.

Universities UK (2010) *Creating Prosperity: The Role of Higher Education in Driving the UK's Creative Economy* (London: Universities UK).

Viale, R. & Pozzali, A. (2010) Complex adaptive systems and the evolutionary Triple Helix, *Critical Sociology*, 36(4), pp. 575–594.

Wood, P. & Taylor, C. (2004) Big ideas for a small town: The Huddersfield creative town initiative, *Local Economy*, 19(4), pp. 380–395.

Wynne, D. & O'Connor, J. (1998) Consumption and the postmodern city, *Urban Studies*, 35(5–6), pp. 841–864.

Spatial–Relational Mapping in Socio-Institutional Perspectives of Innovation

RACHEL C. GRANGER

Department Geography, Environment and Disaster Management, Coventry University, Coventry, UK

ABSTRACT *In recent years innovation studies have extended key discussions beyond scientific knowledge into more symbolic and cultural forms, and with it brought cultural and creative industries to the centre stage of economic innovation. However, the development of more wide-ranging research approaches has failed to keep pace with the advancements occurring in conceptual debates. In this paper, the author draws on the original arguments of the social innovation discourse to highlight the importance of more socialized approaches to the study of innovation, an approach which highlights the importance of understanding more about social networks, local institutions, local scenes and environments, and relational capital. It is known that in practice, and especially with non-scientific knowledge, new ideas are mediated by social relations, institutions and all manner of other intangibles such as conversations; all of which are acknowledged in some literature but have never been part of the mainstream. In this vein, the paper outlines the contribution that spatial–relational mapping can make to the study of innovation by illuminating the social spaces of innovation in and around Coventry and Birmingham, UK, and by raising new patterns of relationships, which emphasize the existence of sector convergence, underground scenes and path lock-in.*

1. Innovation as Drivers of Urban Environments

Increasingly, innovation is viewed as an economic imperative. The act of introducing something new, whether as a commodity or process, has become a central tenet of achieving economic growth, and also for improving quality of life in the face of new global challenges. Businesses, policy-makers and civic authorities look to innovation models to develop new ideas and new technology, improve services and to accelerate this bringing of new knowledge to the market as part of a knowledge-based economic strategy. In a knowledge-driven economy, the focus for many areas is to move up the value chain into more specialized high value-added goods and services and in resourcing research and knowledge transfer to accelerate this. For this reason, the main emphasis of innovation has been one of scientific push, with low levels of R&D cited as reasons for low innovation and low growth—e.g. the low competitiveness and growth of Europe comparative to other global regions (SAPIR *et al.*, 2003). In this environment, the strategy of many urban areas has been to reposition themselves in the global hierarchy, as places rich in skills and technology, and as places where research, innovation and economic growth are self-reinforcing.

One aspect of this strategy which invites quick wins and wider recognition has been the re-designation of industrial areas into creative regions (see discussion by Bontje & Musterd, 2009). In the UK the benefits of hosting creative industries has been in evidence from as early as 1988 (Myerscough, 1988) and part of mainstream economic policy since 1998 (DCMS, 1998). Flagship titles such as City- or Capital of Culture, with their origins in the European initiative add weight to the bold reimagining strategies of several post-industrial cities, now tied to a vibrant creative economy. The European Union's programme on European Creative Districts goes further in linking explicitly a creative economy with innovation and the transformation of local/regional economies (CEC, 2012). They reflect the ambitions of Europe as outlined in the Creative Industries Green Paper, itself influenced by the economic goals of the Lisbon Treaty, which acknowledges "the potential of culture as a catalyst of creativity and innovation" (CEC, 2010, p. 4). The new agenda for the sector (in EU terms) is identified as entrepreneurship and intellectual property, with considerable attention paid to the role of technology in driving such growth. These and other aspects are encompassed in the 2010 Amsterdam Declaration, a set of European-wide policy recommendations to maximize the creative industries as an economic asset across Europe.

Whilst the definition of creative industries is itself the subject of much debate (see discussions by Galloway & Dunlop, 2007; KEA, 2006) the meaning of innovation is more complex still. There remains considerable variation in precisely what innovation means in different fields as Baregheh *et al.* (2009) go some way to uncovering, and the precise form that innovation policy and economic growth initiatives should take is further contested. Over the last two decades the European Union has taken great strides in leading debate on the merits of linear and complex innovation policies, the introduction of territorial policies such as "regional innovation systems", "clusters" and "learning regions", (Cooke & Leydesdorff, 2006), and more recently the debate over what value should be placed on scientific and technological knowledge/innovation over more creative and symbolic forms (see distinction between analytic, synthetic and symbolic knowledge bases made by Asheim *et al.*, 2005). It is not the intention to discuss these aspects here other than for the purpose of contextualizing discussion on research methodologies for innovation studies. In this paper, innovation performance in all its varied forms is recognized as being inherently difficult to deal with, let alone to measure. The speed at which creative activities have been embedded in innovation discourse further adds to this complexity. In this context, the purpose of this paper is to highlight the apparent disconnect between studies on innovation and the realities of innovation, and also the benefits of more relational or socialized approaches to researching this area of activity. This is because increasingly we find that economic activities and formations "originate from social and cultural formations" (Brennan-Horley, 2007). The central role of socialization in creativity and knowledge (see Polanyi, 1967), innovation (see Nonaka & Takeuchi, 1995), and in creating a culture or ecology for creativity and innovation (Bathelt *et al.*, 2004) is now well established, and this provides the starting point for this paper in seeking more imaginative and socially based analytical methods to capture the realities of social-based creativity and innovation activities.

Thus far, methods used for measuring economic success, such as innovation or economic growth, have drawn extensively from accounting disciplines and accordingly tend towards quantifying activities and placing monetary values on processes and outputs. Whilst this has proven useful in making comparative analyses of manufacturing-based economies, this has a more limited utility in conveying the tacit knowledge, symbolic knowledge bases and value embedded in intellectual property that have come to occupy contemporary economies. The increasing importance of qualitative content and flows in the economy highlights the futility of these monetary-based values and quantitative analytical approaches. The reliance on innovation metrics such as R&D expenditure and patents (such as those in the European Innovation Scoreboard) need to be complemented by a broader range of research approaches, which seek to capture real innovation performance in all its varied including qualitative forms. In this sense, intangible and tacit factors can be every bit as important in driving innovation as fixed and tangible aspects of R&D, scientific skills and research capabilities, lending support for more socially and qualitatively nuanced research in this area. This is a recurring theme of this paper, which is structured as three key sections. The first outlines the key areas of the innovation literature and the increasing importance of social and relational accounts of innovation processes. The second section outlines the added value of relational approaches to innovation research, and describes briefly the methodological approach of spatial–relational mapping before examining visual maps. The final section provides an analysis and highlights how relational mapping and associated social network models can provide a more nuanced view of innovation at work in industries and locales.

2. Social Relations and Social Spaces in Innovation

Research on innovation has been invaluable in opening up new avenues of debate over the last two decades but also has been constrained by research methodology. As has been highlighted, there are limitations in how far accountancy-based research methods can be transposed to new areas of economic activity. There is also evidence of dominance of scientific and hi-tech case studies in innovation studies, which has narrowed policy interest to high-value scientific and technological industries, most notably, pharmaceuticals, biotechnology and ICT. This is demonstrated to some degree in Fagerberg and Verspagen's (2009) study of the main areas of innovation literature, in which he cites the important authors, meeting places (e.g. DRUID, EARIE) and journals in the field (e.g. *Research Policy, Economics of Innovation and New Technology, Journal of Evolutionary Economics*), collectively dominated by scientific and technological studies. It is evident from a review of the literature over the last decade using some of Fagerberg's key authors and journals that the number of studies and research papers devoted to analytic or synthetic knowledge bases (science-based discoveries and engineering developments) and the science, technology and innovation (STI) mode of innovation outweighs those studies of symbolic knowledge bases (art and culture) and the doing, using and interacting (DUI) mode of innovation. It is not that analytic or synthetic knowledge bases or STI modes are any more innovative than symbolic knowledge bases or represent a larger section of the economy, but arguably lend themselves to existing quantitative research methods.

Yet while much is now known about the way R&D can spill over from science institutions across actors to generate increasing economic value through innovation (as the STI mode), this is not universal and can operate in vastly different ways, depending on local knowledge and industrial make-up. It is also considerably different to the DUI mode of innovation. Second, the current approach to researching innovation using technology indicators such as R&D, patents and formal innovation partnerships underestimates the extent of innovation in any given area; a point the OECD (2005) concurs with in its assessment of the extent of innovation in the UK, relative to other countries. One of the UK's strengths is in knowledge-intensive services, especially creative industries, in which innovation has been fundamentally underestimated using existing indicators (DIUS, 2008). In these areas, technology indicators and accountancy methods have underplayed the more important issues of problem-solving, know-how and know-who capabilities, semiotic knowledge, networks and other context-specific factors which are important in creating synthetic and symbolic knowledge (e.g. niche manufacturing, creative industries). The overall effect has been one of portraying economic knowledge, activities and value within an undersocialized framework, which acknowledges the role of human action but fails to capture this in any convincing or meaningful way. In other words, it fails to capture the importance of social relations as a type of relational capital. While there has been a growing acceptance of relational capital in some disciplines, this is not manifest in mainstream approaches or research studies in innovation fields.

In new institutional economics, the role of social relations has been taken more seriously as a process through which activities are shaped and knowledge mediated, but has never been incorporated into empirical models. In the New Economic Geography (NEG), which examines uneven spatial development by considering economic processes of agglomeration and dispersal, relational capital, social networks, local scenes and social spaces—as spatiality—play an important role (see overview of NEG by Garretsen & Martin, 2010). This has enabled constructs such as "socialization", "local spillovers", "local tacit knowledge", "propinquity", "clusters" and "local ecologies" to be developed within a wider orthodoxy, to explain the agglomerations of knowledge activities and innovations. And yet even in these areas, accounts have remained largely conceptual or, where attempted, empirical studies have been predominantly economic in nature. It is argued here that this provides a simplistic account of the innovation process, which underplays the importance of local institutions and contexts, and relational social capital such as trust and reciprocity, which in practice can form powerful operating frameworks that shape the trajectories of economies.

3. Socio-Institutional Perspectives

3.1 *Social Spaces*

The need to capture more social and intangible aspects of innovation as demonstrated above is not a new argument; it was expressed in early accounts of "social innovation" discourse during the 1960s. Social innovation today has developed into a set of principles with a strong association with emancipatory politics and the social economy but is an area from which we can learn much about innovation studies.

While social innovation remains an ambiguous concept, many of the analytical lines of enquiry have developed beyond critical responses to oversimplistic (and undersocialized) accounts of the economy and urban change/planning, to arguments about the social form and transformation of society and economy, with an acknowledgement of the social operating mechanism, which drives this. In these different accounts, social innovation is used to denote: (i) *Socialized innovation*—in which social relations are viewed as a key strategic ingredient of innovation processes (see Moulaert *et al.*, 2005), and emphasize the importance of social capital networks in developing economic competitiveness. (ii) *Community Development*—in which social relations are central to integrated area development initiatives (Moulaert *et al.*, 2005) with a reference to the way in which they can lead to improvements in social or community life and to changes in the relations between local groups and individuals Nussbaumer and Moulaert (2004). (iii) *Social economy*—reflecting innovations in the non-profit or third sector, with commercial innovation objectives being used for social good or for wider social concerns. (iv) *Social-global governance*—reflecting changes in social attitudes and cultures, culminating in some instances in key social movements.

Notwithstanding these differences in definition and orientation, social innovation is drawn upon as a framework for research despite its ambiguous and contested meaning because of the way emphasis is placed on social relations and social context or capital. Social relations and the richness imbued in social capital both provide a basis for cultivating spatial, economic and political perspectives of social reality, an idea which draws on sociological research such as the early and seminal work of Granovetter (1973). While in some cases then, social innovation could be interpreted as "innovation for societal progress" and relates to the idea of changes that improve beliefs, basic routines and authority flows in society, it also speaks of the notion of "social capital", which is an important component of socialized accounts of innovation. It connotes the idea of resources accruing from pools of social relations as well as highlighting that networks imbued with different relations not only have value in yielding resources but are also powerful constructs through which one understands social-led processes.

In the different accounts of social innovation, (social) relations are emphasized as key ingredients and also the principal vectors for change, so that innovation is shown to reflect the spatial representation and expression of different relations, institutions and interactions, which combine to produce changes in approaches and systems (Ditts & Westley, 2009). Institutions here could be viewed as frameworks of norms, and rules and practices, which structure relations and action in social contexts and reflect the rhythms of daily life (Giddens, 1984). They are therefore consonant with a broader "relational turn" in the social sciences (see discussion Bathelt & Glückler, 2002, 2003).

3.2 *Spatial–Relational Mapping*

Drawing on sociological institutionalism to emphasize the importance of different social institutions and social network analysis to emphasize social relations, "spatial–relational mapping" is presented here as a tool for examining how and where economic knowledge is mobilized across actors and geographical settings. As such it provides a broad methodological approach for examining knowledge-driven processes occurring through relational ties, which offers a more realistic account of the relational networks, which shape creativity, innovation and the spatial economy. In other words, it conveys the social infrastructure or ecology through which creativity is mediated and innovation occurs, and in doing so captures the integral relationship between economic and spatial form.

In practical terms, spatial–relational mapping provides a tool for examining some of the more ethereal aspects of NEG, which have thus far been explored in largely conceptual ways. "Buzz" (Bathelt *et al.*, 2002, 2004), "noise and signals" (Grabher, 2002), "chatter" (Jones, 2007) and "information ecologies" (Grabher, 2004) describe local and relational features of an operating framework, which are active. They capture the idea of active information and learning networks said to materialize from face-to-face contact within an industry at a given point in time, which produce high levels of action and interdependency, and a social capital, which cannot be replicated in other forms. While in much of the literature this has been treated conceptually rather than empirically, in the few empirical studies that exist, researchers have tended to fall short and expose structural rather than relational capital because of the difficulty of capturing these ethereal forms within conventional research approaches. For example, Mould and Joel's (2010) portrayal of formal partnership boards and the composition of these succeeds in drawing attention to individuals as agents of knowledge transfer, but not the active conversations, interdependencies and buzz that shape London's advertising industry in practice. In Brennan-Horley and Gibson's (2009) attempt to capture relational origins of creativity, there are practical difficulties in conveying rich ethnographic accounts through GIS mapping techniques.

In a point of departure then, spatial–relational maps as outlined here depict active and empirical relationships and conversations through nodes (actors) and relational ties within a spatial context. This enables researchers to view how knowledge is mobilized in practice. They show in map form, which interactions penetrate and shape economic activity; which locations and people (or organizations) act as glue or platforms in economies, and who and what drives these. As such, it offers a richer snapshot of the "buzz" and "chatter" of industries in practice and the networks that underpin these, while helping to explain the fine grain of knowledge activity occurring and embedded in specific locales. In practice, they have the potential to answer the questions about where the creative heart or "epicentre" of a city really is (Brennan-Horley & Gibson, 2009). Whilst spatial–relational maps draw heavily on earlier notions of actor network theory (Latour, 2005) and uses of social network analysis, they bring both economic activities (relational capital) and geo-spatial contexts together to show activities in a spatial as well as relational manner.

In the remainder of this paper, spatial–relational maps of the UK's West Midlands Performing and Visual Arts sector are presented and analysed. The area includes Birmingham, the second city in England and encompasses a highly rich symbolic knowledge industry. Whilst the maps were constructed solely from survey material and were not complemented by other social surveying techniques, they are presented here not as exhaustive research approaches but as additional insights that might address the current shortcomings in mainstream methodologies for capturing relational assets and social networks.

4. Mapping Underground Scenes, Networks and Lock-in

During 2009 and again in 2010, the author began to examine the creative economy through spatial relational mapping, and drew on the West Midlands economy as a local case study. During this research a binary matrix of self-disclosed ties between different actors in the same or in associated industries were constructed. Actors were asked during survey to identify relational ties important for creativity and innovation. From this, a series of maps were produced, which presented individuals and organizations as a series of nodes, and their "ego networks" as a set of mini networks or platforms that drive creativity and innovation in their industry and area. The case study used was the West Midlands visual and performing arts sector in and around Coventry and Birmingham, in which the author was based.

Coventry is a medium-sized city in the West Midlands, UK, and has a well-established Arts Sector comprising visual art and media sub-sectors, dance, music, theatre and performing arts. The region's capital city is Birmingham, which also acts as the country's second city. Both cities lie 20 miles apart. Visual and performing arts have a proud tradition in Coventry, stemming from the influential arts-based activity of Coventry University[1] and the Herbert Art Gallery and Museum, the reputation of the Warwick Arts Centre and Belgrade Theatre based in Coventry, and organizations based around three key locations in the city: the ICE on the technology Park; the arts studios and warehouses located at the Canal Basin; and the Far Gosford Street Creative Village, Fargo. In Birmingham, the visual and performing arts sector encompasses digital media sub-sectors, in which Birmingham specializes, and is served by several higher and further education institutions, and established creative industry communities in and around the Digbeth area of the city.

Using a small number of key contact points in Coventry ($n = 3$) as a starting point for a snowballing technique, contacts were asked to name important people they have regular contact with (to exchange ideas, to source information, to trade with and also network with) who are important in driving creativity and innovation. Participants were asked to measure the strength or importance of these ties and also to identify groups or networks they occupy, which bring them into contact with other people in and outside of the industry. From the results, it is possible to discern whether some actors and network spaces are more important than others in driving creativity and innovation in the arts in the West Midlands.

The approach has permitted a series of relational maps to be constructed showing the incidence and breadth of active relations between people and organizations in the industry, the extent of personal and shared networks (expressed as ego networks), and the significance of key people, places and events. All of which have been expressed in a geographical context to give added meaning to where the "noise", "chatter" and "buzz" of the sector is most frequently located.

4.1 Spatial and Sectoral Convergence

What the research and maps uncover is a strong convergence between sub-sectors, which in other research approaches have tended to be treated separately, as separate sector studies. In the West Midlands, while a number of workers operate within distinct sub-sectors such as poetry or playwriting, there is also evidence of cross-over, the extent of which has been underplayed previously in economic research. For example, one social media consultant was found to work frequently within visual artists, musicians, journalists and even glass makers, and featured in several sub-networks, while one dancer found contact with visual artists and poets highly important for her work. Also found was a large number of creative workers operating in multiple roles up- and down-stream of value chains or across sub-sectors ($n = 65$). In one such example, an artist operated as a gallery manager, a consultant, university lecturer and musician.

What this reveals is that the spaces occupied by some individuals and the conversations recorded were found to be multiplex, enabling contact between sub-sectors and creating a diverse environment from which ideas can cross-fertilize. However, given that these actors could not be attributed to any one sector (and in some cases to any one geographical space) with any certainty or fixity (see similar discussion by Amin & Cohendet, 2004), the author would raise a methodological point about the need to determine in what capacity participants are revealing relations and social spaces (i.e. in which of their job roles). This crossover of jobs and sub-sectors, which is evident through the maps, which otherwise would be masked by sector or location studies, can be taken as a salutary reminder of applying rigid occupation and industry classifications in the study of creative sub-sectors more generally, in which it is now apparent that individuals operate in multiple capacities and sub-sectors.

4.2 *Underground Creative Spaces*

The research also reveals an especially rich network of opportunities and creative spaces in and around Coventry and Birmingham, which include a formal layer of organizations such as Arts Council England, National Council for Graduate Entrepreneurship and the work of theatres, galleries, universities, local authorities and research institutes. These could be viewed as a highly visible but perhaps formal layer of actors, fixed in space and time and operating within full view of those working in the field.

Below this are a set of less formal but visible spaces, linked to organizations but argued to be more fluid and social in nature. While these remain highly visible to the sector, they operate within less formal frameworks, have less stringent norms of behaviour, operate under the principle of open access and are not attributed to any one location. As such, they operate in multiple places or at multiple times. These spaces include semi-formal and semi-organized partnerships, groups and communities of practice such as the former Coventry Artist Network (now defunct), Coventry's Emerge Network, JEEcamp a self-help group for journalists and the Red Teapot network.

Building on the ideas of Cohendet *et al.* (2009), it is possible to distinguish between these formal and semi-formal meeting spaces outlined above, which could be termed, respectively, the upperground and middleground, with more informal and frequently hidden spaces, which might represent an "underground" of activity. These include the varied spaces offered by online blogs and communities of practice such as social media, informal cafes, bars and clubs as well as popular social meeting places such as festivals (predominantly in Birmingham, Figure 1) or the shop-front theatre spaces in Coventry. The recently developed Shop-Front theatre, which operates from disused retail spaces in Coventry's precinct, appears to be an underground space in so far as it is informal and transient, but acts as an important glue for attracting creative workers, and platform for different parts of the industry. Different dancers, visual artists, actors/directors, playwrights etc. converge at the temporary theatre spaces run by Theatre Absolute and Art Space principally to socialize but facilitating wider professional discussions and activities. The creative activities of the more recent FarGO Village project, which is host to different creative events, are also gaining a reputation as a hub of activity in Coventry.

At the same time, several participants noted the importance of Birmingham's online blogs, communities of practice and platforms such as "Creative Republic", "Paradise Circus" and "Made in Birmingham". To this, one might add the Birmingham Social Media Café, which acts as a visible but also informal space for digital media workers, Polar Bear the record shop, Thursday nights at the Prince of Wales pub and Rooty Tooty in Birmingham's Digbeth/Custard Factory area. These represent recognized

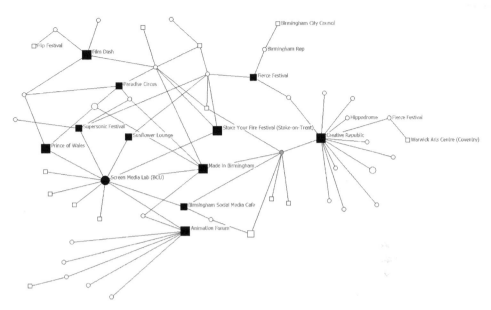

Figure 1. The importance of festivals as platforms for relational capital.

spaces operating within and across sub-sectors, which are known to creative workers but which are not formally recognized as local cultural assets. In other words they operate out-of-view from many practitioners and policy-makers occupying more formal spaces and networks, but nevertheless act as the glue in the local sector.

Middle-ground spaces include the Herbert Cafe in Coventry (In the Herbert Art Gallery and Museum), Artspace that offers studios and meeting spaces for visual artists in Coventry, the semi-formal meetings of the Emerge Network, which bring different members of the industry together, and also the Animation Forum in Birmingham.

4.3 *Organizational and Sector Lock-in*

In looking at the individual transactions that make up the visual and performing arts sector, the mapping exercise reveals a degree of separation occurring between the public and private sector, which might denote poor interoperability between different parts of the industry. This is reflected in the positive correlation between the source sector of a participant and the sector s/he networks with. As Table 1 shows, of the 177 relations recorded from participants operating in the public sector, 111 (65%) occurred with other public-based contacts. Similar patterns are recorded between private-to-private and freelance-

Table 1. Interactions among Coventry's creative industries, by sector

	Public/upper	Private/middle	Freelance/underground
Public	111	51	15
Private	60	114	15
Freelance	18	24	36

to-freelance interactions, in a process, which ultimately militates cross-sector chatter and cross-fertilization of ideas.

Moreover, the research revealed a positive correlation between sector and spatial scale, so that public sector workers were found largely to occupy upperground spaces such as formal partnerships based in organizations, the Belgrade Theatre, Warwick Arts Centre, the Herbert Art Gallery and others. Conversely, freelance workers and micro-businesses were found to interact with and communicate almost exclusively with other freelancer workers and businesses via online communities of practice, bars and festivals. Thus, Birmingham's comparatively stronger presence of underground scenes (comparative to Coventry) could be argued to offer a richer ecology of information and learning networks—or buzz—than that of Coventry, in which the public sector or upperground dominates and which might be said to be more institutionally thin. The degree to which this might create a lock-in of capital within an area is a highly significant prospect, which is suggested in the maps but has not thus far been considered in different studies of innovation in creative sectors.

4.4 *Lock-in*

Lock-in has been used previously in an institutional or business context (Grabher, 1993) to denote the insular and uncompetitive nature of networks, as the benefits conferred from a network's existing strong ties begin to restrict new thinking and innovation. In innovation and regional development discourse, lock-in has been used to denote a situation where technology evolves in a certain way or follows a certain path because of the increasing returns between a set of actors, which self-reinforce the network and force out alternatives (Cowan & Gunby, 1996). In many respects, this can create an embeddedness dilemma, in which a network or group has become strong but begins to mark the limits of effective networking, thus posing a dilemma for policy-makers and planners. While lock-in then has been acknowledged as being important in producing a space for engagement and for building a system of innovation, it is used here to refer to the pejorative effects of trapping in and locking skills and ideas.

In Coventry, a strong network of information exchange, networking and trading was identified among a group of small theatre companies, an arts-based organization and key public-sector organizations, all operating within the city ring road. These seven organizations acknowledge the strength of their internal relationship and this is denoted by the strength of lines used between nodes in Figure 2. Also of note is the insular nature of the network in which there are no obvious visible external links with other actors, and which collectively suggest the beginning of a locked-in network of relational capital. There is supplementary evidence to support this picture through accounts of these actors drawing on the forged links within the network to source information, to tap into groups of freelance artists, to inter-trade and generally to garner support for new projects. As one key stakeholder notes "they inter-trade extensively . . . but are less alert to creating a profile and acting strategically". This raises a pertinent question for many urban areas, which have attempted to construct advantage through the support of new creative sub-sectors: at what point does a network with strong embedded links cease to be vibrant and become a closed network that displays lock-in? Second, at what point might a town or city plan to intervene in an organic sector/network to prevent lock-in occurring, thereby adding to the public sector/upperground dominance

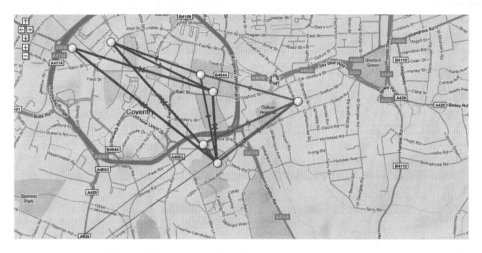

Figure 2. Spatial–relational lock-in, Coventry (extract).

and diluting chatter? These are issues that have been raised through the mapping approaches.

5. Conclusions

What has been argued in this paper is that to understand the practice of innovation, one needs to understand the relational capital and social spaces through and by which innovation occurs. The exclusive use of economic accounting models, conceptual models or empirical models that focus on analytic knowledge bases and science and technology can tell us only so much. Finding ways of capturing relational capital in empirical accounts of innovation, which can be used in different sectors, can potentially tell us much more about who drives innovation, where this occurs and which social spaces act as glue and platforms.

Against this backdrop, the use of spatial–relational mapping, which draws on the relational approaches of actor network theory and social network analyses, and combines with geographic mapping (or GIS) can open up different views of an area or industry. In the study of the West Midlands Visual and Performing Arts sector, the spatial–relational mapping has highlighted the emergence and importance of: (i) sector convergence and the idea that individuals and organizations operate in multiple sectors and multiple places; (ii) the existence of upper-, middle- and under-ground spaces or scenes, which operate almost exclusively and effectively separate and isolate groups within the industry; (iii) the importance of some underground spaces tied to social functions, e.g. festivals, informal theatre spaces, cafes, which act as platforms for bringing people together and effectively act as glue that binds the sector; (iv) path lock-in or the idea that strong ties create powerful networks, which drive innovation but become too strong over time and lead to diminishing innovation. The maps reveal new patterns, which could be subjected to further testing and which would not be identified through economic, sector or location studies. The maps also provide a more convincing approach, "spatially literate" about locating creative vibes or centres of cities/economies, which many researchers and policy-makers are at pains to fix in space. As such spatial–relational mapping might have immediate utility for planners and policy-makers hoping to understand more about a given area or sector, or the creative and innovation processes that underpin these.

Second, the orthodoxy of NEG as referenced in the second section of this paper underscores the role of face-to-face tacit knowledge exchange, untraded interdependencies between local actors and the power of information and learning ecologies as building blocks of innovation. These different aspects demand more nuanced sociological investigation, a call which has been answered here to some extent through the application of spatial–relational mapping. It is not argued that spatial–relational mapping should replace conventional approaches to innovation studies, especially those based on accounting methods. This is because the maps reveal only so much. Whilst they depict the functional activities of a selected area and industry, by showing networks and ties, and gathering points, ultimately they cannot depict the (economic) value of some ties more than others in economic growth, which would be better served through economic and survey methods. What can be gleaned from the approach however are broad patterns occurring in a sector such as the emergence of bottle necks, gaps in contact and insular activities, which prompt policy intervention as well as area's interoperability—how much local actors are operating together—and the degree of interdependencies, which in economic and conceptual studies have been argued to be building blocks of innovation. In other words, the maps address an existing weakness in innovation research by offering a more socialized framework, and paint a different picture of an area or sector, which might complement other methods. As such, whilst the approach is not without critique—the technique might not be said to be statistically robust—it certainly offers an alternative approach, which might be tested in new case studies and, in so doing, open up new avenues of discussion.

Note

1. The visual art, performing arts and media teaching activities of Coventry School of Art and Design, established in the nineteenth century, and the research activities of the Institute for Creative Enterprise (ICE), Serious Games Institute, an applied research institute for the gaming and software industries, and the Design Institute, a specialist research centre for all aspects of design

References

Amin, A. & Cohendet, P. (2004) *Architectures of Knowledge. Firms, Capabilities and Communities* (Oxford: Oxford University Press).

Asheim, B. T., Coenen, L., Moodysson, J. & Vang-Lauridsen, J. (2005) Regional innovation system policy: A knowledge-based approach. CIRCLE Paper 2005/13, Lund, Sweden: Lund University.

Baregheh, A., Rowley, J. & Sambrook, S. (2009) Towards a multidisciplinary definition of innovation, *Management Decision*, 47(8), pp. 1323–1339.

Bathelt, H. & Glückler, J. (2002) *Wirtschaftgeographie: Ökonomische Beziehungen in Räumlicher Perspective* [Economic Geography: Economic Relations in Spatial Perspective] (Stuttgart: UTB-Ulmer).

Bathelt, H. & Glückler, J. (2003) Toward a relational economic geography, *Journal of Economic Geography*, 3(2), pp. 117–144.

Bathelt, H., Malmberg, A. & Maskell, P. (2002) Cluster and knowledge: Local buzz, global pipelines and the process of knowledge creation. DRUID paper 02/12, Danish Research Unit for Industrial Dynamics, Copenhagen, Denmark: Copenhagan Business School.

Bathelt, H., Malmberg, A. & Maskell, P. (2004) Clusters and knowledge: Local buzz, global pipelines and the process of knowledge creation, *Progress in Human Geography*, 28(1), pp. 31–56.

Bontje, M. & Musterd, S. (2009) Creative industries, creative class and competitiveness: Expert opinions critically approved, *Geoforum*, 40(5), pp. 843–852.

Brennan-Horley, C. (2007) Work and play: Vagaries surrounding contemporary cultural production in Sydney's dance music culture, *Media International Australia*, 123, pp. 123–137.

Brennan-Horley, C. & Gibson, C. (2009) Where is creativity in the city? Integrating qualitative and GIS methods, *Environment and Planning A*, 41(11), pp. 2595–2614.

CEC (2010) *Unlocking the Potential of Cultural and Creative Industries,* COM(2010) 183 (Strasbourg: Commission of the European Communities).

CEC (2012) *European Creative Districts*, 34/G/ENT/PPA/12/6483 (Strasbourg: Commission of the European Communities).

Cohendet, P., Grandadam, D. & Simon, L. (2009) Places, spaces and the dynamics of creativity. International Conference on Organizational Learning, Knowledge and Capabilities (OLKC), Amsterdam, 26–28 April.

Cooke, P. & Leydesdorff, L. (2006) Regional development in the knowledge-based economy: The construction of advantage, *The Journal of Technology Transfer*, 31(1), pp. 5–15.

Cowan, R. & Gunby, P. (1996) Sprayed to death: Path dependence, lock-in and pest control strategies, *The Economic Journal*, 106(436), pp. 521–542.

DCMS (1998) *The Creative Industries Mapping Document 1998. Dept. Culture, Media, and Sports* (London: TSO).

Ditts, A. & Westley, F. (2009) *Complexity and Convergence: How Intertwining Roles and Interests Create Conditions for the Success of a Social Innovation. Centre for Social Innovation* (Waterloo, ON: University of Waterloo).

DIUS (2008) *Innovation Nation. Department for Innovation, University and Skills, UK* (London: TSO).

Fagerberg, J. & Verspagen, B. (2009) Innovation studies – the emerging structure of a new scientific field, *Research Policy*, 38(2), pp. 218–233.

Galloway, S. & Dunlop, S. (2007) A critique of definitions of the cultural and creative industries in public policy, *International Journal of Cultural Policy*, 13(1), pp. 17–31.

Garretsen, H. & Martin, R. (2010) Rethinking (new) economic geography M: Taking geography and history more seriously, *Spatial Economic Analysis*, 5(2), pp. 127–160.

Giddens, A. (1984) *The Constitution of Society* (Cambridge: Policy Press).

Grabher, G. (1993) The weakness of strong ties: The lock-in of regional development in the Ruhr area, in: G. Grabher (Ed.) *The Embedded Firm: On the Socioeconomics of Industrial Networks*, pp. 255–277 (London: Routledge).

Grabher, G. (2002) The project ecology of advertising: Tasks, talents and teams, in: G. Grabher (Ed.) *Production in Projects: Economic Geographies of Temporary Collaboration, Regional Studies Special Issue*, 36(3), pp. 245–63 (London: Routledge/Taylor and Francis).

Grabher, G. (2004) Learning in projects, remembering in networks? Communality, sociality and connectivity in project ecologies, *European Urban and Regional Studies*, 11(2), pp. 103–123.

Granovetter, M. S. (1973) The strength of weak ties, *American Journal of Sociology*, 78(6), pp. 1360–1380.

Jones, A. (2007) More than "managing across borders"? The complex role of face-to-face interaction in globalizing law firms, *Journal of Economic Geography*, 7(2), pp. 223–246.

KEA (2006) The economy of culture in Europe. KEA European affairs. Report produced for DG Education and Culture, Strasbourg: Commission of the European Communities.

Latour, B. (2005) *Reassembling the Social: An Introduction to Actor-Network-Theory* (Oxford: Oxford University Press).

Moulaert, F., Martinelli, F., Swyngedouw, E. & Gonzales, S. (2005) Towards alternative model(s) of local innovation, *Urban Studies*, 42(11), pp. 1969–1990.

Mould, O. & Joel, S. (2010) Knowledge networks of "Buzz" in London's advertising industry: A social network analysis approach, *Area*, 42(3), pp. 281–292.

Myerscough, J. (1988) *The Economic Importance of the Arts in Glasgow* (London: Policy Studies Institute).

Nonaka, I. & Takeuchi, H. (1995) *The Knowledge Creating Company* (Oxford: Oxford University Press).

Nussbaumer, J. & Moulaert, F. (2004) Integrated area development and social innovation in European cities: A cultural focus, *City*, 8(2), pp. 249–257.

OECD (2005) *Economic Surveys: UK* (Paris: OECD).

Polanyi, M. (1967) *The Tacit Dimension* (New York: Anchor Books).

Sapir, A., Aghion, P., Bertola, G., Hellwig, M., Pisani-Ferry, J., Rosati, D., Viñals, J. & Wallace, H. (2003) An Agenda for a growing Europe. Making the EU economic system deliver. Report for the European Commission, Strasbourg: Commission of the European Communities.

Creative City Policy and the Gap with Theory

JAN JACOB TRIP & ARIE ROMEIN

OTB Research Institute for the Built Environment, Delft University of Technology, Delft, The Netherlands

ABSTRACT *The creative city concept is popular among researchers and policy-makers. On the one hand, academic literature elaborates, on a conceptual level, the importance of creativity and innovation for urban competitiveness; on the other, numerous cities develop and implement creative city policies in practice. The connection between these two is rather weak and, accordingly, creative city policy tends to be ad hoc. Our purpose in this paper is to narrow the above-mentioned gap between theory and practice, by addressing the question of how conceptual insights into the creative city can be converted into an elaborated operational approach for local policy practice. We propose a three-step approach: (1) to position a city's current creative places and communities within the context of social and economic structures, urban narratives and prevailing governance structures and style by means of a systematic analytical framework; (2) to assess the spatial, social and symbolic place qualities of the creative production and consumption; (3) to identify options for effective policy intervention. We further examine how these steps may be applied in practice, and use the city of Delft in the Netherlands as an example. A discussion of the applicability and implementation of this approach concludes the paper.*

1. Introduction

In recent years the creative city concept has become highly popular among local policy-makers and researchers. Particularly in the wake of Florida's book *The Rise of the Creative Class* (2002) and Landry's *The Creative City* (2000), every self-respecting city seems to have turned its attention to developing a creative city policy. Even though the creative city concept generally refers to larger cities that can provide a certain critical mass and diversity of people and activities, small and medium-sized cities and regions also advertise themselves as "creative" and set out creative city policies (e.g. INTELI, 2011), seemingly without sufficient awareness of the limitations of creative city policies. Furthermore, the creative city thesis—particularly the work of Florida—quickly became the subject of fierce academic debate (e.g. Peck, 2005; Zimmerman, 2008; Atkinson & Easthope, 2009; Long, 2009).

Furthermore, a gap exists between the academic literature which discusses the development of the creative city or the creative economy on a conceptual level, and actual policy development in individual cities. Many cities base their creative city policies explicitly on a limited number of sources that make more or less empirically based academic insights accessible to a broader audience. These include, notably, the above-mentioned works of Landry and Florida. Furthermore, cities are apt to imitate—with questionable chances for success—well-known creative city success stories such as Barcelona or Lille (cf. Brenner, 2003; Harris, 2006). Both practices carry the risk that local policy-makers "fall into a reductive trap of universality at the cost of understanding the particular" (Evans, 2009, p. 1006). Finally, some policy-makers pay lip service to the creative city concept, considering it a label to "renew" and popularize existing policies (Chatterton, 2000, p. 392). As Russo and van der Borg (2010, p. 686) state, the relation between culture and urban economic development remains largely "a black box in which most cities move like amateurs". Accordingly, creative city policy tends to be ad hoc, or, as Jayne comments (in: Evans, 2009, p. 1011) with regard to the implementation of a creative industry agenda at the regional level in the UK, "[. . .] at best patchy . . . a lack of best-practice models and empirical research to guide policy-makers".

Our purpose in this paper is to narrow the gap between theory and practice. We address the question of how conceptual insights into the creative city can be converted into an operational approach for policy practice that is tailored to the local context. This is informed by studies on creative city policy in a number of Dutch and European cities (Romein & Trip, 2009a, 2009b, 2010; Trip & Romein, 2009). We focus on policy related to the qualities of urban space, as this is considered a main factor defining the success of creative city policy. We do not propose a creative city blueprint as we consider this probably the least appropriate approach with regard to creative city development. Nevertheless, we contend that an approach based on the ideas presented in this paper may contribute to a comprehensive, city-specific and focused creative city policy.

The focus of the paper is therefore methodical. First we discuss some conceptual aspects of the creative city and creative city policy that constitute the starting points of our analysis. In Section 3, we outline the proposed method itself. In Section 4, we apply the framework developed in the previous sections to assess the state of the creative economy in the Dutch city of Delft as an illustration of how the described methodology could be applied, rather than as an in-depth case study. A discussion of the applicability and implementation of the proposed approach concludes the paper.

2. Conceptual Background

Within the academic debate on creativity-based economic development various approaches coexist which focus on cultural or creative industries, on a broad "creative class" or on creative cities (see, e.g. UNDP/UNCTAD, 2010 for an extended overview). Furthermore, a large part of the debate focuses on the demarcation of creative activities, occupations, people or economic sectors. As, for example, Evans (2009, pp. 1020–1022) shows, a variety of definitions is applied that makes it almost impossible to draw unambiguous comparisons between cities, especially on an international level. A discussion of these wide-ranging debates is beyond the scope of this paper. Instead, we focus on the most relevant aspects for our aim to narrow the gap between theory and policy.

2.1 *Creative Production and Consumption Milieu*

If we ignore for a moment the myriad of conceptual differences and divisions within the current debate it may be concluded that two perspectives on the creative economy prevail both in academic literature and policy practice:

(1) a *production milieu* or business-oriented approach, focusing on the role of creative industries as generators of employment and innovation. This perspective considers the creative industries as a regular economic sector, although with some rather specific characteristics: small, but nevertheless significant for the innovativeness of the urban economy, not least due to its relations with conventional firms, and to a large extent based on small firms and face-to-face contacts. Scott (2000), for instance, analyses the functioning of the fashion, furniture and film industries in Los Angeles and Paris. Similarly, Kloosterman (2010) focuses on the architecture sector in Rotterdam and Amsterdam, and Currid (2007) on the role of fashion, art and music in New York. Others, such as Pratt (2004) and Lorentzen and Frederiksen (2008), focus on creative clusters from a more general perspective;

(2) a *consumption milieu* or people-oriented approach, emphasizing the role of qualities of cities in attracting creative talent. This perspective gained importance in the academic debate only in recent years. It starts from the assumption that "jobs follow people" or "labour follows capital", which is opposite to the traditional view. Cities should therefore attract creative talent and—the bottom line—businesses will follow. The most renowned advocate of this approach is Florida (2002), who based his creative class concept largely on the "jobs follow people" assumption. Related to this, albeit less explicitly focusing on creative people, is research that focuses on the role of urban amenities in economic development (e.g. Glaeser *et al.*, 2001; Clark, 2004). With hindsight, the work of Jane Jacobs (1961) may also be connected to this approach, as she emphasized many characteristics of what is now understood as the creative city.

These two perspectives are largely complementary. The creative industries are characterized by a relatively large number of small firms, self-employed and freelance workers, who often work from home or choose a business location close to where they live. This implies that working, living and leisure very much intertwine for creative workers, and that "consumption"-related amenities such as shops and schools, but also, for instance, the visual quality of an area, partly affect the location of businesses (cf. Smit, 2011). Moreover, the "buzz" upon which many creative activities depend (Storper & Venables, 2002) and which is emphasized in the production milieu approach is typically rooted in a specific local context which consists for a large part of factors that are generally related to the consumption approach. This may include certain pubs and venues functioning as "third spaces", but also liveliness and cultural amenities (cf. Grabher, 2002; Currid, 2007). Hence, a "creative city" should combine a production and a consumption milieu attractive to the creative industries, and the policies supporting creative industries have to take both approaches into account (cf. Russo & van der Borg, 2010, p. 681).

In spite of the above, policy practice tends to focus mostly on the production milieu, particularly on creative businesses and entrepreneurs (cf. Banks & O'Connor, 2009, p. 386). Policies regarding the consumption milieu usually have a more general focus on residential quality.

Several factors may contribute to the prevalence of the production milieu approach in creative city policy. A main reason is that the primary motives to develop a creative city policy tend to be economic rather than cultural. Many cities turned to the creative economy to compensate for the loss of employment in manufacturing and standardized services—which relocated to low-cost environments—rather than out of altruistic motives such as art promotion. Consequently, creative city policy is to a large extent entrepreneurial policy. Nonetheless, creative industries do not fit into the traditional economic model and a creative city policy that focuses mainly on growth and profit therefore "produces many conflicts and confusions, especially at the local level, where enthused policy-makers confront a sector often sceptical or simply unable to act in the expected manner of a dynamic, emergent 'growth sector'. Policy-makers, then, need to understand these complex dimensions, though it is generally agreed that this has rarely been the case" (Banks & O'Connor, 2009, p. 386; cf. Russo & van der Borg, 2010, p. 671).

One of the complicating features of creative industries that contributes to an explanation of the above observations is the diversity of entrepreneurs' motives and ambitions. On the one hand, creative industries as they are commonly defined in current literature include architects, graphic designers, copywriters and photographers who manage to combine a distinct cultural profile whilst operating on a commercial base. On the other hand, however, many visual artists, for example, attach great importance to cultural integrity and creative originality and are uncomfortable with the label "entrepreneur". These tend to disregard commercialization of their work for fear of damaging their creative integrity (cf. Rutten *et al.*, 2005; Flemming, 2007; Rae, 2007; Jacobs, 2009). Conflicts may arise when policy-makers focus on economic development and fail to recognize the specific character and broader requirements of creative industries and creative workers: the need for affordable housing and working spaces compared with the gentrification of "creative" neighbourhoods; the day-to-day cultural life of creatives compared with cultural-touristic mega-events; or low-profile culture venues compared with landmark museum buildings and concert halls. As a result, creatives of both the "artistic" and the "commercial" type occasionally together protest against creative city policy. Recent examples are Rotterdam, where creatives published a plea for a less elitist and growth-driven "uncreative city" (BAVO, 2007), and Hamburg, where creative entrepreneurs protested that local authorities abused their reputation for city marketing purposes and to commercialize the city (Die Zeit, 2009).

Rather than disqualifying the production milieu approach per se, as Banks and O'Connor (2009) suggest, these examples once more indicate that creative city policy needs to address both the production and consumption milieu as a unity rather than separately. Moreover, it needs to be clearly focused rather than targeting a wide range of creative industries, entrepreneurs and workers whose interests differ or may even conflict with each other in practice.

2.2 *Place Qualities*

The above discussion leaves the question of what an attractive, creative production and consumption milieu consists of. Florida (2002) emphasizes the concept "quality of place", including a range of qualities stretching from bike tracks, meeting places and popular music venues to authenticity and tolerance. Others similarly stress the importance of "intangibles" for the creative production and consumption milieu (Kotkin, 2000; Landry, 2000; Florida & Gates, 2004; Trip, 2007). However, elusive qualities such as authenticity, identity and openness are hard to define and even more difficult to measure and plan effectively. Moreover, as, for instance, Kotkin (2000) and Florida (2008) show, creative workers are not a homogeneous category; they have different preferences according to age, stage of life and personal attitude and circumstances, which can hardly be addressed by generic policies.

Silver *et al.* (2011a, 2011b) go even further as they propose a theory of "scenes", looking at the specific meaning of coherent sets of amenities and urban assets for specific social groups. Their work is not specifically aimed at creative people, let alone at the creative economy. Nevertheless, their focus on the meaning of amenities is crucial. Urban space is neither an empty medium nor a neutral category in which "things" are localized and activities take place. Instead, it is both an expression of social identities and relationships, and a medium that creates and reproduces these identities and relationships. In other words, "things" and activities are not just there, but have a significance, which is different for different creative people.

Place qualities are most commonly emphasized in the consumption milieu approach. Nevertheless, they are also relevant from the production milieu perspective, particularly regarding the role and functioning of creative clusters (cf. CURDS, 2001; Pratt, 2004; Scott, 2006). In general, the existence of clusters of firms is explained by cost reduction due to two types of agglomeration economies external to the firms: localization and urbanization economies. Localization economies derive from the co-location and collective action of firms in the same type of industry whereas urbanization economies result from the interrelations between firms in different and unrelated industries. Sources of localization economies are forward and backward supplies, information exchange and knowledge spill-overs, a pool of specialized labour and the availability of a range of auxiliary trades and specialized services (e.g. finance, vocational training and research institutes). Urbanization economies are "regarded as a function of the scope and diversity of production within the urban concentration" (Parr, 2002, p. 160).

Both types of external economies also apply to creative industries (Pratt, 2004; Bagwell, 2008). Lorentzen and Freriksen (2008, p. 159) indicate specialization versus the diversity of industries, labour, institutions and infrastructure as the key differences between both types of external economies with regard to creative clusters. Their characterization of urbanization economies as hinging "upon a range of place specific and idiographic factors that are invariably urban" is more telling. These factors include qualities of place, such as housing, cultural and recreational amenities, tolerance towards diverse lifestyles and "buzz" (Storper & Venables, 2002).

The production and consumption milieu are defined by a complex and unique set of urban qualities. To unravel these, we distinguish three types of urban space: (1) "social space" involves the network of functional relationships and social interactions; (2) "symbolic space" represents the perception of the specific significance of places by the people who use them; and (3) "physical space" includes both the morphology and the location pattern of urban functions, including their accessibility.

These three types of space are dynamic; they change under the influence of societal processes, each according to its own particular pattern. Over time an increasing stratification may be observed, particularly of social space, as activity patterns become more divergent in scope (reflecting the increased individualization and diversification of life styles) and scale (reflecting the increasing spatial range of activities). Furthermore, changes in each of these three spaces may influence the other two: the places we go to and the people we meet there affect our own and other people's appreciation of those places, but even our appreciation of other places, as they also affect our reference framework.

The creative production and consumption milieu in a city or neighbourhood are defined by place qualities that may be related to these three types of space. Thus, for instance, housing, working spaces, infrastructure and amenities may be considered physical place qualities; relation networks, meeting places and street life social place qualities; and image and authenticity symbolic place qualities. Because of their increasing stratification, these different spatial qualities are appreciated by creative entrepreneurs and workers on different geographical scales, particularly the city-region, city or district within the city. Nevertheless, it is frequently observed that creative clusters tend to concentrate within centrally located urban districts of larger cities that offer a distinct, favourable combination of place qualities (cf. Kloosterman, 2004; Markusen *et al.*, 2008).

Many of the above-mentioned place qualities can hardly be planned for, especially not in the short term. This implies that the success of a city as a creative city partly depends on the existence of a certain potential. Regarding the consumption milieu, this indicates an advantage for cities with preserved historic neighbourhoods, many cultural amenities and, for instance, universities—the kind of environment where a creative consumption milieu is "in the air" and that is hard to reproduce. It also suggests an advantage, in general, for larger cities. For the production milieu, a critical mass of creative activities and people seems the best starting point for further development of the creative economy. Creative city policy might successfully exploit these factors, but except for in the very long term can hardly create them. This leads to the conclusion that creative city policy should build "organically" on what is already there, rather than on a tabula rasa. This, in turn, means it should be city-specific rather than based on duplication of best practices, or, in the words of Pratt (2011, p. 123), "situated and not universal".

2.3 *Analytical Framework*

The various spatial-physical, social and symbolic place qualities that define the creative production and consumption milieu are summarized in an analytical framework of the creative production and consumption milieu (Table 1). The schedule as presented here is far from exhaustive, but provides a good impression of the range of place qualities that are relevant to the development of the creative economy. These have been based on an inventory of factors mentioned in both the academic and semi-academic literature on creative city development (Trip & Romein, 2009, pp. 218–220).

As Table 1 shows, many place qualities refer to either the production or the consumption milieu. Nevertheless, several also refer to both types of milieu, reflecting the close relations between working and living among workers and entrepreneurs in creative industries.

The variety of place qualities that are relevant to the development of the creative production and consumption milieu indicates that creative city policy cannot be limited to a single policy field. As Russo and van der Borg (2010, p. 687) state, policy with respect to the creative economy needs "to span all government operations, rather than being con-

Table 1. Analytical framework of the creative production and consumption milieu

	Production milieu	Consumption milieu
Physical space	Quality and price level of working spaces Availability of business services Presence of knowledge-intensive industries Research and education infrastructure Concentrations of businesses (creative clusters)	Quality and price level of housing Availability of household amenities (child day care, schools, shops, sport, etc.)
	Diversity and density of built environment "Quality architecture" Availability of combined working/living spaces Availability of amenities	
Social space	Relation networks (within creative industries and between creative industries and other sectors) Creative meeting places ("third places") Presence of "buzz" Diverse labour pool	Diversity of people Diversity of jobs
	Liveliness, "street life" (Tolerance of) cultural diversity (Tolerance of) social diversity	
Symbolic space	Authentic cultural heritage "Tale" or "DNA" of the area Sense of community "Creative" image	

fined to 'cultural departments'", or, we may add, to economic departments. It should be integrative, involving such different policy fields as economic and social policy, spatial planning, housing, culture and leisure, as well as policy culture itself.

3. Three-Step Approach

This section briefly elaborates on the elements of a comprehensive policy approach to narrow the gap between academic discussion and policy practice. It entails three steps, with a Strengths, weaknesses, opportunities, threats (SWOT) analysis confrontation type at its heart. The discussion in Section 2 provides the basis in two ways. First, a number of observations will be derived that constitute the conceptual starting points of our approach:

(1) the creative city is taken as a whole rather than as separate production and consumption milieus;
(2) the policy perspective is integrative rather than sectorial, including all relevant policy fields;
(3) policy is city-specific, building "organically" on what is already there.

Second, the discussion above provides the analytical framework for Table 1, which summarizes the way we envisage the creative city.

The three-step approach was applied, by way of example, to analyse the creative economy in the Dutch city of Delft (Romein & Trip, 2009a). Section 4 elaborates on this. Nevertheless, our insights are not based on this single case alone. The method has also been applied to nine cities in six countries independently of the analysis of Delft, within the framework of the Interreg IVB project *Creative City Challenge* (Romein & Trip, 2010). In the latter case, the different context of the project required slight adjustments, as data were obtained from local expert knowledge rather than interviews. Nonetheless, the same basic three steps could be followed, demonstrating the versatility of the method.

3.1 *Step 1: Positioning of the City, Selection of Criteria and Data Collection*

The first part of Step 1 would be to position a city's present creative places and creative communities in the context of social and economic structures, urban narratives and prevailing governance structures and style. Does a (potentially) vital creative economy exist, or is there no more than the beginning of a creative city development present? What is the composition of present creative industries—are they concentrated in, for example, design or games development, or does no particular specialism prevail? Local policy culture is likely to influence how the creative economy is regarded, and how the above questions are answered: the dominant political colour of city government over time, the influence of vested interests or the lobbying capacity of new interest groups.

Answers to questions such as these are crucial to defining the key issue of this first step: what should be the focus of creative city policy? The analytical framework presented in Table 1 may be helpful in this phase, as it provides an overview of the issues that may be relevant.

The analytical framework also provides a concrete starting point for the analysis of the existing creative production and consumption milieu. The factors in Table 1 can be converted into operational indicators for which data can be collected. The number of relevant place qualities may be extended (far) beyond those listed in Table 1 depending on the uniqueness of the city, although in practice it is usually necessary to make a selection. This selection should cover all relevant aspects of the creative production and consumption milieu but should also include some introspection on local policy culture. Any selection should include the place qualities that are regarded as most relevant in terms of the defined policy focus.

In view of the place qualities mentioned in Table 1, any selection of indicators is likely to include both quantitative and qualitative indicators. Quantitative data may be available in particular for physical place qualities such as the supply and price level of housing and working spaces, or for the size and composition of the existing creative sector. Many other place qualities, especially the social and symbolic ones, can only be assessed using qualitative data derived from documents, in-depth interviews or local expert knowledge.

3.2 *Step 2: Assessment of Strengths, Weaknesses, Opportunities and Threats*

In Step 2, the collected data are assessed, in order to designate place qualities as strengths or weakness. Strengths and weaknesses focus on the current situation: "what is there?" As such, they may be considered relatively static. The assessment itself can best be based on local expert knowledge, particularly on interviews with local creative entrepreneurs. The analysis below of Delft is primarily based on interviews, whereas in the case of the *Creative City Challenge* project assessment was done based on the local expert judgement of partners in the project.

Furthermore, Step 2 includes the identification of opportunities and threats. Compared with strengths and weaknesses, opportunities and threats are more dynamic: "what is happening?" They concern processes and developments that possibly influence the creative production and consumption milieu and the chances for creative industry growth. Opportunities and threats may be city-specific, for example, local urban regeneration policy that includes the large-scale demolishment of dwellings or the construction of roads that improve accessibility. Often, however, they are of a general nature, such as national or EU regulations, the growing diversity of lifestyles and its impact on demand for products of the creative industry, or the possible effects of the current financial turmoil and economic downturn.

3.3 *Step 3: Identification of Promising Fields for Policy Intervention*

The result of Step 2 is an inventory of strengths and weaknesses, opportunities and threats for the city involved, i.e. the "four lists" that typically emerge from a SWOT analysis. In Step 3, a confrontation matrix (Figure 1) is applied that goes beyond these lists in order to identify promising fields for policy intervention (cf. Kearns, 1992, p. 13). The strengths and weaknesses, opportunities and threats that were identified in Step 2 are ranked along the margins of the matrix. The quadrants of the matrix indicate the four possible combinations, each of which has other implications for policy. Four types of policy options can be formulated: (1) "invest" in promising strengths to exploit comparative advantages; (2) mobilize resources to "defend" threatened strengths; (3) "decide" whether to invest to strengthen promising, but weak areas; (4) "control damage" caused by weak and threatened areas by avoiding them and looking for alternatives. In practice not all combinations of strengths or weaknesses and opportunities or threats result in useful policy options; only those which have a logical "match". A match is more probable if both factors concern the same type of (production or consumption) milieu. Eventually, the detailed policy options in the quadrants of the confrontation matrix may be aggregated to distinguish a limited number of comprehensive fields that appear promising for effective policy intervention regarding the development of the creative economy.

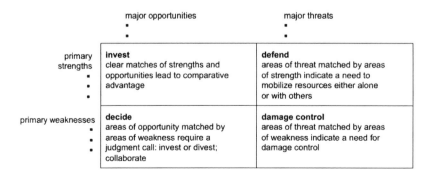

Figure 1. The confrontation matrix.
Source: Kearns (1992, p. 13).

The next section outlines the application of this approach by means of an analysis of the creative production and consumption milieu in the city of Delft.

4. Delft as a Creative City

This section presents Delft as a case study of a creative city. With almost 97,000 inhabitants (Statistics Netherlands, 1-1-2010) Delft is a medium-sized city within the Dutch urban system. It illustrates that not only large cities set out a creative city policy. However, rather than being an in-depth study of the creative and cultural industries of Delft, the three steps described above are used to analyse the city's production and consumption milieu, in line with the overall aim of this paper.

The analysis of Delft started with desk-based research of the main recent policy documents and local and regional statistics regarding the city's creative economy. These were based on the classification of creative industries as defined by TNO (Dutch Organization for Applied Science). Since its introduction in 2004 (Rutten *et al.*, 2004), this classification is the standard for analysing the creative industry in the Netherlands (Manshanden & van Bree, 2010). It distinguishes three sectors within the creative industries, and a number of links in the cultural value chain (Figure 2). Within this value chain two demarcations of the creative industries can be distinguished. The "narrow" definition was applied here, which includes creation, production and exploitation of cultural goods, while the "broad" definition also entails distribution activities (e.g. book stores). Each cell in Figure 2 represents a sector and value chain combination that corresponds to a series of specific SBI/ISIC branch codes (Braams & Urlings, 2010, p. 4).

Based on information obtained from the desk-based research, a format was designed for semi-structured in-depth interviews with a number of non-randomly selected creative entrepreneurs. These interviews aimed to provide a clear general understanding of entrepreneurs' assessment of the current production and consumption milieu of Delft, reflecting the analytical framework of Table 1 as well as entrepreneurs' opinion about possible opportunities and threats for Delft as a creative city.

We interviewed 11 entrepreneurs who were spread as evenly as possible over the subsectors of the creative industry in Delft and these subsectors' geographic distribution across the various creative milieus of the city (Table 2). These lengthy in-depth interviews provided ample information about the motivations that were behind creative entrepreneurs'

sectors / value chain	arts and heritage	media and entertainment	creative business services
creation			
production			
publishing and exploitation			
distribution			
consumption			

Figure 2. Value chain of creative industries according to the TNO classification; the dark area represents the "narrow" definition applied here.
Source: Rutten *et al.* (2004, p. 26).

Table 2. Interviewees according to creative subsector and location

	Inner city	University campus	Elsewhere	Total
Arts	3			3
Media and entertainment			2	2
Creative producer services	3	2	1	6
Total	6	2	3	11

assessment of place qualities. Both the selection of the interviewees—branch and location of entreprenuers—and the outcomes of the interviews were discussed with local policy-makers in the municipal departments of Economy and Culture, the latter in order to validate certain observations and conclusions.

In addition to the interviews, information was obtained from discussions with representatives of creative industries branch organizations and knowledge institutions. Furthermore, an expert meeting was organized on the role of place qualities in the creative production and consumption milieu in the Delft region, in which all of the above-mentioned parties were represented, as well as representatives of cities comparable to Delft, such as Leiden, Groningen and Eindhoven.[1] This rich contextual information obtained from the additional discussions and expert meeting enabled the researchers to obtain a sufficiently complete picture of the creative production and consumption milieu of Delft from a relatively limited number of interviews.

4.1 *Step 1: Positioning of the City, Selection of Criteria and Data Collection*

Delft is located in the densely urbanized western part of the Netherlands, between Rotterdam and the Hague, both of which can reached within about 15 minutes by train, and with populations of 593,000 and 489,000 inhabitants, respectively (Statistics Netherlands, 1-1-2010) are considerably larger than Delft. In recent decades Delft has been severely hit by de-industrialization, particularly in the food-processing industry (Fernández-Maldonado & Romein, 2008).

Two features of Delft are of particular relevance to the city's creative production and consumption milieu. First, the city is home to several major research and education institutes, notably Delft University of Technology (DUT) and the Netherlands Organization for Applied Scientific Research (TNO). Both are located south of the inner city, on and adjacent to a university campus, respectively. DUT has about 4600 employees and 17,000 students (DUT, 2011), making it the largest university of technology in the Netherlands. TNO has several locations in Delft, together accommodating almost 1200 employees (figure acquired by e-mail). Second, Delft has a historic inner city that dates back to the Middle Ages which is lively as well as picturesque. This inner city is an important element of the city's creative production and consumption milieu, as a location for creative firms, a meeting place and a podium for cultural activities; however, it is also expensive, to a great extent preserved, and to some people too much of an "open-air museum".

Creative industries in Delft provide over 2200 jobs, 4.7% of total employment (Table 3). Creative producer services are by far the largest subsector, accounting for over 60% of the jobs and almost half of the firms in the creative sector. Creative producer services in Delft include a relatively large number of architects and designers. This can be related

Table 3. Creative industries in Delft by subsector (1 January 2008)

	Employment		Business structure	
	Jobs	Per cent of total employment	Firms	Jobs per firm
Arts	576	1.2	153	3.8
Media and entertainment	256	0.5	108	2.4
Creative producer services	1381	2.9	255	5.4
Total creative industries	2213	4.7	516	4.3
Total Delft	47,299	100.0	3134	15.1

Source: figures obtained by e-mail from the Hague Region (2009).

directly to the university's faculties of Architecture and Industrial Design. They are among the largest faculties in Delft, and the largest of their kind in the Netherlands, and teach a range of creative disciplines. In the case of Architecture this ranges from design to building technology to urbanism, while Industrial Design involves, for instance, automotive design. Many of the city's creative entrepreneurs and workers studied here and still maintain close relations with the university by giving guest lectures, offering apprenticeships and recruiting employees. Many internationally renowned Dutch architects hold or have held special chairs in the faculty of Architecture, such as Kees Kaan (Claus and Kaan Architects), Pi de Bruijn and Carel Weeber (Architecten Cie.) and Francine Houben (Mecanoo). Conversely, many professors whose primary tasks are in the faculty are also active as architects or urban development consultants one or two days a week. The relation between chairs and design practitioners is less explicit in the faculty of Industrial Design, but several active designers from Delft hold or have held chairs in the institution, as have several former designers (from, e.g. Philips and furniture company Gispen).

Other branches of the creative industries, although of lesser importance in Delft, are also present. While a large part of the interviewees were architects, designers and ITC developers, an artist, a children's book author, an organizer of pop concerts and the manager of a local popular music venue were interviewed as well. Some of these also have connections to the university and other knowledge institutes, for instance, by giving courses or workshops.

The major clusters of creative industries in Delft are its historic inner city and the university campus on the south side of the city plus the adjacent DelftTech business park. Overall, the arts subsector is typically located in the inner city, whereas ICT and construction technology firms are primarily located on the university campus due to a strong relation with the technological faculties of the university. The selection of entrepreneurs for interviews reflects this spatial spread of subsectors (see Table 2).

4.2 *Step 2: Assessment of Strengths, Weaknesses, Opportunities and Threats*

Table 4 presents creative entrepreneurs' assessment of place qualities as strengths and weaknesses. Besides these opposite assessments, all qualities are also assessed as neither strengths nor weakness (neutral) by the entrepreneurs. Although the number of actual weaknesses is small, the ratio between strengths and neutral issues may also

Table 4. Creative entrepreneurs' assessment of Delft as a location for creative industries

	Strength	Neutral	Weakness	Unknown
Physical quality				
Physical space in (inner) city	10	1	–	–
Residential climate and environment	7	4	–	–
Accessibility and parking	3	2	5	1
Supply of working spaces	2	4	5	
Supply of residential spaces	–	8	2	1
Social quality				
Relations to higher-education institutions	8	2	1	–
Formal relation networks of "creatives"	7	4	–	–
Proximity to and cooperation with other creative businesses	5	6	–	–
"Third places"	3	8	–	–
Lively inner city as location for unplanned meetings	3	8	–	–
Social tolerance and openness	3	4	–	4
Symbolic quality				
Solidarity with city's atmosphere and icons	9	2	–	–
Solidarity with creative community	4	7	–	–
Creative city image	2	2	7	–

reveal a lot. Strengths include the historic inner city (for its urban space and atmosphere rather than as a meeting place), the general residential climate and atmosphere, and the relations between creative industries and higher education institutions, particularly the university. Weaknesses can be found in a number of more concrete, practical issues, such as accessibility and parking (in the inner city), and the supply of affordable working space and housing for personnel. Creative entrepreneurs appreciate Delft but do not generally consider it a "creative city". They were less outspoken about other social place qualities, such as the usefulness of formal networks of creative entrepreneurs or "third places". There is no lack of third places as such, but creative entrepreneurs do not use them very often. With regard to cooperation and customer relations, the central location of Delft from a regional perspective was often mentioned as an advantage.

Analysis of these assessments, and the views and considerations behind these as expressed during the interviews, results in the strengths and weaknesses presented in Table 5. A few issues were added that were mentioned repeatedly, but more implicitly in the interviews. In addition, entrepreneurs' were also asked to identify possible opportunities and threats that might affect both types of milieus. As Table 5 shows, relatively few opportunities and threats were indicated. Opportunities include the redevelopment of industrial buildings to lessen the shortage of working spaces for creative industries, and the more open attitude of the university towards relations and cooperation with creative industries. Threats include the difficulty of retaining creative graduates. Remarks were made that the inner city tends to be too much of a "historic museum" in the eyes of young creative talent, and has little to offer for graduates who can no longer depend on student-oriented services and amenities. Furthermore, particularly larger businesses felt that policy focused too much on starting entrepreneurs, rather than addressing, for instance, the problem of business expansion in the protected inner city.

Table 5. Overview of strengths, weaknesses, opportunities and threats regarding Delft as a location for creative industries

Strengths	Weaknesses
• Built environment in inner city	• Insufficient supply of affordable and suitable working spaces
• Solidarity of creative entrepreneurs with city of Delft	• No creative city image
• Availability of formal relation networks	• Insufficient supply of affordable and suitable residential spaces
• Relation to higher-education institutions	• Accessibility by car/congestion due to construction works; parking
• Intake of young creative talent	• "Open-air museum" character of inner city
• Situation between Rotterdam and the Hague	• Few "big players" or "drivers" in creative sector
	• Insufficient spatial quality of university campus
Opportunities	Threats
• Availability of spaces for redevelopment into working spaces for creative industries	• City is unable to retain young urban talent (graduates)
• More open attitude of university towards creative industries	• Policy does not evolve with growth of creative sector and creative businesses

4.3 *Step 3: Identification of Promising Fields for Policy Intervention*

The strengths, weaknesses, opportunities and threats were combined into a confrontation matrix, as described in Section 3 (Table 3). The policy options in the matrix' quadrants can be aggregated into a number of promising fields for policy intervention:

(1) *Retain young creative talent (graduates).* It is difficult for Delft to attract and retain young creative talent, mostly graduates from the Architecture and Industrial Design faculties. These could either start their own business or go to work for an existing firm in the creative sector; however, many leave Delft because of the shortage of affordable working and residential spaces, the lack of cultural and leisure amenities, the poor image of Delft as a creative city and the "open-air museum" character of the inner city. In the longer term this threatens the position of Delft as a creative city. Therefore, it is necessary to pay more attention to the interests of young creative talent in all the relevant policy fields;

(2) *Accommodate the growth of creative businesses.* In the inner city, demand for space by successful and expanding businesses often conflicts with the protected monumental status of buildings. Other problems connected to an inner city location, such as parking and lack of space, also tend to become more important when firms grow. The shortage of large companies as "drivers" of the creative sector in Delft is a weakness for the city; therefore it is important to tie also larger creative firms to Delft and to involve them actively in the development of the creative city;

(3) *Utilize the city's location and guarantee its accessibility.* These issues are not specific for the creative industries, but affect particularly larger creative business, whose location requirements for a large part are similar to those of firms in other sectors. Its central location enables Delft to "borrow size": inhabitants of Delft have relatively

	OPPORTUNITIES (1) Availability of spaces for redevelopment into working spaces for creative industries; (2) More open attitude of university towards creative industries	**THREATS** (3) City is unable to retain Young urban talent (graduates); (4) Policy does not evolve with growth of creative sector and creative businesses
STRENGTHS (A) Built environment in inner city (B) Solidarity of creative entrepreneurs with city of Delft (C) Availability of formal relation networks (D) Relation to higher education institutions (E) Intake of young creative talent (F) Situation between Rotterdam and the Hague	**(I) INVEST** Stimulate redevelopment of working spaces for creative industries, also as a meeting place ("third places") (B/C1) Stimulate participation of university in formal networks (C2) Expand and intensify contacts between university, creative industries and municipality ("triple helix") (D2)	**(III) DEFEND** Accommodate demand for space by creative industries whilst maintaining quality of inner city (A2) Hold on to creative talent in the city by means of atmosphere, icons; increasing identification of young creative talent with city (B/C/E1) Retain specific working space for small/starting creative businesses (C2) Keep up amenities for young creative talent (F1) Maintain possibility to retain larger creative businesses (F2)
WEAKNESSES (G) Insufficient supply of affordable and suitable working spaces (H) No creative city image (I) Insufficient affordable and suitable residential spaces (J) Accessibility by car/congestion due to construction works (K) Parking (L) "Open-air museum" character of inner city (M) Few "big players" or "drivers" in creative sector (N) Insufficient physical quality of university campus	**(II) DECIDE** Stimulate redevelopment of working spaces for creative industries (G1) Stimulate redevelopment of working spaces for creative industries in combination with housing and spaces for combined working and living (I1) Guarantee accessibility of (re)developed locations (J/K1) Improve physical quality of university campus as a location for creative industries (N2)	**(IV) DAMAGE CONTROL** Maintain balance between monumental character of inner city and transformation into working spaces (G/L4) City image not appropriate to retain young creative talent (H2) Young urban talent leaves city due to shortage of suitable affordable housing (I3) Restrictive parking policy a problem for businesses in and around inner city (K4) "Open-air museum" character of inner city hardly interesting for young creative talent (L3) Campus provides insufficient quality as a "third place" (N3)

Figure 3. Confrontation matrix regarding Delft as a location for creative industries

easy access to the larger supplies of amenities in Rotterdam and the Hague. The more relaxed housing market in these cities may to some extent mitigate the shortage of housing for personnel in Delft itself. The accessibility of Delft is a problem in this respect, however. Accessibility by car, using the A13 motorway, is bad for a large part of the day. Accessibility by train is good, but the connection between the station and locations outside the inner city is weak, such as DelftTech south of the university campus where several creative firms are located. This makes both commuting unattractive and the accessibility of firms problematic. Accessibility and parking within Delft are a point of attention for local policy, particularly as a relatively large share of creative businesses are located within the inner city. Furthermore, the municipality, possibly together with the university, should make a case to the regional and national governments to improve the external accessibility of the city;

(4) *Strengthen the relation between the city and the university.* The role of the university in Delft directly involves the "triple helix" of entrepreneurs, government and knowledge institutions that constitutes an important basis for the creative economy. Particularly relevant here are the faculties of Architecture and Industrial Design of DUT. Relations between these faculties and the creative sector in Delft are numerous, and have the potential for further growth. The same is true, to a lesser extent, for the relations between the university and the municipality. Joint initiatives that support graduates starting their own business, such as the Yes! Delft incubator for techno-starters and the planning and masterplan for the Technological Innovation Campus Delft, make it possible to further strengthen the triple helix. Other initiatives, in addition, could make the university campus a more attractive business location and meeting place for creative entrepreneurs.

5. Discussion

In the previous sections we proposed a three-step approach to narrow the gap between academic debate and policy practice with respect to creative city development. Its application was demonstrated by means of an illustrative case study of the Dutch city of Delft. The analysis indicates that all three types of place qualities—social, symbolic and physical—prove to be relevant for the creative city policy of Delft. Also, the creative production and consumption milieu in Delft are strongly related; for instance, the available housing stock affects the ability of firms to retain creative workers or the attractive historic inner city poses problems for creative firms that wish to expand. In addition to giving insight into the creative dimension of Delft, the results of the analysis prove that our three-step methodology is accurate. It actually confirms a number of local policy-makers' assumptions that so far remained largely latent and unnamed, and shows how these can be transferred into practical policy interventions in a diversity of policy fields.

Specific creative city policies in many cities tend to focus on the production milieu, whereas policies aimed at the consumption milieu often have a more general focus on "quality of life". Indeed, most cities' creative city policies were induced by economic motives in the first place, and economic departments are often among the first to adopt the ideas of Florida and other creative city champions. A business-oriented approach to creative city policy is not necessarily a problem, as long as policy also takes into account the distinct characteristics of the creative industries. Evidence from Delft supports the idea that creative firms increasingly resemble "normal" firms when they grow and professionalize, and "traditional" location factors such as accessibility, parking and opportunities (including space) for business expansion become more important. Nevertheless, these larger firms also share distinct preferences for certain place qualities that characterize the creative industries. A creative milieu is also for them broader than production alone; it involves a creative atmosphere and image of the city in which producers, consumers, knowledge institutes and intermediaries like trend watchers or "tastemakers" (Currid, 2007, p. 108) in specific creative industries are interconnected.

In practice, therefore, a strong focus on economic policy aims may indeed raise problems. First, it often means that creative city policy is pursued by economic departments with little participation by the housing, spatial planning, culture and social policy departments. However, the coherence of the creative production and consumption milieu implies that these departments have to cooperate in a pragmatic way, on the basis of a shared, coherent vision. Second, a focus on economic development increases the risk of a creative city policy that is strongly "data-driven", leading to an undervaluation of social and symbolic place qualities which can hardly be grasped in clear-cut statistics. Finally, the plea for an integrated approach also concerns economic policy as such. A creative city policy that is restricted to the growth of some defined creative branches per se offers limited prospects for economic performance in the end, since it is unlikely that creative industries as such will amount to more than, say, 5–10% of total employment. It is essential, therefore, to strive for close connections between creative industries and other, presumably knowledge-intensive, industries, both in terms of policy and on a business-to-business level.

Assuming that manufacturing in the old Fordist sense can no longer provide a firm economic base for most western high-cost cities, creative city policy is often considered a welcome answer to industrial decline or even a paradigm shift of how cities currently grow and develop (e.g. Bontje et al., 2011). Nevertheless, the limits of creative city policy should be acknowledged as well. Creative city development partly depends on a city's past development. Many important place qualities—not least the intangibles that are now emphasized by creative city protagonists, such as authenticity and tolerance—can hardly be planned for in the short term. Instead, they evolve in the course of time, usually in the long term. A degree of path dependency is involved, meaning that the range of possibilities for present creative city polices is limited by economic and policy trends, decisions and events from the past. This concerns concrete characteristics such as a city's built environment and economic structure, but also its institutions and policy culture. Thus, creative city development, as a relatively new perspective, may have to confront "frozen" mindsets, policy conventions and vested interests.

Current creative city policy-makers, then, should be realistic in a dual sense: starting from the place qualities and policy culture "that are there" rather than blindly copying best practices from other cities, and applying a long-term vision rather than planning for short-term successes (cf. Brown & Meczynski, 2009). The latter implies that policies should not give priority to short-term expectations or profits, even if the private actors involved are generally assumed to aim for short-term efficiency, profit and shareholder value. Neither should policy-makers cease their efforts with the first sign of criticism or crisis. On the other hand, creative city planning policy has to avoid a rigid top-down approach and blueprint type. Instead, it should exercise restraint and put into practice a type of planning that aims at incremental development.

The call for realism also once again raises the issue of scale. The creative city concept generally refers to medium-sized and larger cities that can provide a diverse range of place qualities and a sufficient critical mass of relevant people, activities and locations. Nevertheless, this may not necessarily mean the creative city policies pursued by many small and medium-sized cities will ultimately be futile, if these smaller cities are able to "borrow size". Delft has some distinct strengths in its design sector, its knowledge institutions and its historical inner city, but at best is considered a medium-sized city. The lack of scale may partly be solved by the explicit embeddedness of creative city policy within an urban region (Rotterdam–the Hague) with a broader potential. The question is, then, whether policy-makers in smaller cities are able to adopt a supra-local perspective, and whether they can accept the limits of their city's potential for creative city development.

6. Conclusion

The discussion above indicates that effective creative city policy is no easy policy. Creative city development requires an integral approach by both the production and consumption milieu that surpasses the boundaries between various policy fields and departments of local government; it must be based on a long-term vision that is resistant to economic setbacks and the pressure of excessive short-term expectations; and it has to be city-specific, taking into account existing assets, shortcomings, vested interests and policy conventions. Cities that just surf the hype largely depend on good luck and are likely to be disappointed, unless there already was a flourishing creative economy.

However, the bottom line to this must be that creative city policy cannot be the responsibility of public authorities alone. Rather, creative city development requires collective action by government, entrepreneurs—creative and other types—and knowledge institutions, together often referred to as the triple helix (Etzkowitz, 2008). The examples of Delft and other cities indicate that this triple helix functions in practice and that cooperation within the triple helix, based on mutual trust and understanding, is a crucial precondition for successful creative city policy. This is most likely to grow over time and requires continuous efforts as creative city policy evolves.

Acknowledgement

The research on which this paper has been based was carried out partly within the North Sea Region Interreg IVB project *Creative City Challenge*.

Note

1. Expert meeting on "Place qualities: the 'forgotten' dimension of the agenda of the Rotterdam-Delft knowledge region?", 27 May 2011, Delft.

References

Atkinson, R. & Easthope, H. (2009) The consequences of the creative class: The pursuit of creativity strategies in Australia's cities, *International Journal of Urban and Regional Research*, 33(1), pp. 64–79.

Bagwell, S. (2008) Creative clusters and city growth, *Creative Industries Journal*, 1(1), pp. 31–46.

Banks, M. & O'Connor, J. (2009) Introduction: After the creative industries, *International Journal of Cultural Policy*, 15(4), pp. 365–373.

BAVO (2007) Pleidooi voor een oncreatieve stad [Plea for an uncreative city], *Metropolis M*, 28(1), pp. 27–31.

Bontje, M., Musterd, S. & Pelzer, P. (2011) *Inventive City Regions. Path Dependency and Creative Knowledge Strategies* (Farnham: Ashgate).

Braams, N. & Urlings, N. (2010) *Creatieve industrie in Nederland; creatieve bedrijven* [Creative industries in the Netherlands; creative firms] (The Hague: Statistics Netherlands).

Brenner, N. (2003) Stereotypes, archetypes, and prototypes: Three uses of superlatives in contemporary urban studies, *City and Community*, 2(3), pp. 205–216.

Brown, J. & Meczynski, M. (2009) 'Complexcities': Locational choices of creative knowledge workers, *Built Environment*, 35(2), pp. 238–252.

Centre for Urban and Regional Development Studies (CURDS) (2001) *Culture Cluster Mapping and Analysis* (University of Newcastle Upon Tyne). Available at http://ncl.ac.uk/curds (accessed November 2009).

Chatterton, P. (2000) Will the real creative city please stand up? *City*, 4(3), pp. 390–397.

Clark, T. N. (2004) Urban amenities: Lakes, opera, and juice bars: Do they drive development? in: T. N. Clark (Ed.) *The City as an Entertainment Machine*, pp. 103–140 (Amsterdam: Elsevier).

Currid, E. (2007) *The Warhol Economy: How Fashion, Art, and Music Drive New York City* (Princeton, NJ: Princeton University Press).

Delft University of Technology (DUT) (2011) *TU Delft Highlights*. Available at https://epub01.publitas.nl/84/42/magazine.php - spreadview/40/ (accessed January 2012).

Die Zeit. (2009) "Lasst den Scheiß!" Ein Künstler-Manifest gegen die Hamburger Kulturpolitik ["Cut the crap!" An artists' manifesto against Hamburg's culture policy]. *Die Zeit*, 7 November 2009. Available at http://www.zeit.de (accessed January 2010).

Etzkowitz, H. (2008) *The Triple Helix; University-Industry-Government Innovation in Action* (New York: Routledge).

Evans, G. (2009) Creative cities, creative spaces and urban policy, *Urban Studies*, 46(5/6), pp. 1003–1040.

Fernández-Maldonado, A. M. & Romein, A. (2008) A knowledge-based urban paradox: The case of Delft, in: T. Yigitcanlar, K. Velibeyoglu & S. Baum (Eds) *Knowledge-Based Urban Development. Planning and Applications in the Information Era*, pp. 221–240 (New York: Information Science Reference).

Flemming, T. (2007) Investment and funding for creative enterprises in the UK, in: C. Henry (Ed.) *Entrepreneurship in the Creative Industries*, pp. 107–125 (Cheltenham: Edward Elgar).

Florida, R. (2002) *The Rise of the Creative Class: And How It's Transforming Work, Leisure, Community and Everyday Life* (New York: Basic Books).

Florida, R. (2008) *Who's Your City? How the Creative Economy Is Making where to Live the Most Important Decision of Your Life* (New York: Basic Books).

Florida, R. & Gates, G. (2004) Technology and tolerance: The importance of diversity to high-technology growth, in: T. N. Clark (Ed.) *The City as an Entertainment Machine*, pp. 199–217 (Amsterdam: Elsevier).

Glaeser, E. L., Kolko, J. & Saiz, A. (2001) Consumer city, *Journal of Economic Geography*, 1(1), pp. 27–50.

Grabher, G. (2002) The project ecology of advertising: Tasks, talents and teams, *Regional Studies*, 36(3), pp. 245–262.

Harris, H. (2006) Supercity me; on contextualising the creative city discourse within contemporary urban studies. Paper presented at the International Conference 'Urban Conditions and Life Chances', 6–8 July 2006, Amsterdam.

INTELI (2011) *Creative-Based Strategies in Small and Medium-Sized Cities: Guidelines for Local Authorities* (Lisbon: URBACT II Programme of the European Commission).

Jacobs, J. (1961) *The Death and Life of Great American Cities* (New York: Random House).

Jacobs, D. (2009) A creative city is more than a city of creatives [in Dutch], in: S. Franke & G. J. Hospers (Eds) *The Living City. About the Contemporary Meaning of Jane Jacobs* [in Dutch], pp. 57–67 (Amsterdam: SUN Trancity).

Kearns, K. P. (1992) From comparative advantage to damage control: Clarifying strategic issues using SWOT analysis, *Nonprofit Management & Leadership*, 3(1), pp. 3–22.

Kloosterman, R. C. (2004) Recent employment trends in the cultural industries in Amsterdam, Rotterdam, the Hague and Utrecht: A first exploration, *Tijdschrift voor Economische en Sociale Geografie*, 95(2), pp. 243–252.

Kloosterman, R. C. (2010) Building a career: Labour practices and cluster reproduction in Dutch Architectural design, *Regional Studies*, 44(7), pp. 859–871.

Kotkin, J. (2000) *The New Geography* (New York: Random House).

Landry, C. (2000) *The Creative City* (London: Earthscan).

Long, J. (2009) Sustaining creativity in the creative archetype: The case of Austin, *Cities*, 26, pp. 210–219.

Lorentzen, M. & Frederiksen, L. (2008) Why do cultural industries cluster? Localization, urbanization, products and projects, in: P. Cooke & L. Lazzaretti (Eds) *Creative Cities, Cultural Clusters and Local Economic Development*, pp. 155–182 (Cheltenham: Edward Elgar).

Manshanden, W. & van Bree, Th. (2010) *De Rotterdamse creatieve industrie: indicatoren van een stedelijk cluster* [The Rotterdam Creative Industry: Indicators of an Urban Cluster] (Delft: TNO Bouw en Ondergrond).

Markusen, A., Wassall, G. H., DeNatale, D. & Cohen, R. (2008) Defining the creative economy: Industry and occupational approaches, *Economic Development Quarterly*, 22(1), pp. 24–45.

Parr, J. (2002) Missing elements in the analysis of agglomeration economies, *International Regional Science Review*, 25(2), pp. 151–168.

Peck, J. (2005) Struggling with the creative class, *International Journal of Urban and Regional Research*, 29(4), pp. 740–770.

Pratt, A. C. (2004) Creative clusters: Towards the governance of the creative industries production system? *Media International Australia Incorporating Culture and Policy*, 112, pp. 50–66.

Pratt, A. C. (2011) The cultural contradictions of the creative city, *Culture, City and Society*, 2(3), pp. 123–130.

Rae, D. (2007) Creative industries in the UK: Cultural diffusion or discontinuity? in: C. Henry (Ed.) *Entrepreneurship in the Creative Industries*, pp. 54–71 (Cheltenham: Edward Elgar).

Romein, A. & Trip, J. J. (2009a) , *Nulmeting monitor creatieve economie Delft* [Baseline Measurement Creative Economy Delft] (Delft: Delft University of Technology). Available at http://repository.tudelft.nl (accessed February 2012).

Romein, A. & Trip, J. J. (2009b) *Theory and Practice of the Creative City Thesis: The Cases of Amsterdam and Rotterdam.* PNDonline, I-2009. Available at www.planung-neu-denken.de (accessed February 2012).

Romein, A. & Trip, J. J. (2010) *The Creative Economy in CCC Cities and Regions*, SWOT Analysis Report 6.2, written within the framework of the NSR INTERREG IVB project Creative City Challenge. Delft: Delft University of Technology. Available at www.creative-city-challenge.org (accessed February 2012).

Russo, A. P. & van der Borg, J. (2010) An urban policy framework for culture-oriented economic development: Lessons from the Netherlands, *Urban Geography*, 31(5), pp. 668–690.

Rutten, P., Manshanden, W., Muskens, J. & Koops, O. (2004) , *De creatieve industrie in Amsterdam en de regio* [The Creative Industry in Amsterdam and the Region] (Delft: TNO Informatie en Communicatietechnologie).

Rutten, P., IJdens, T., Jacobs, D. & Koch, K. (2005), *Knelpunten in creatieve productie: creatieve industrie* [Bottlenecks in Creative Production: Creative Industries] (Delft: TNO Informatie en Communicatietechnologie).

Scott, A. J. (2000) *The Cultural Economy of Cities; Essays on the Geography of Image-Producing Industries* (London: Sage).

Scott, A. J. (2006) Creative cities: Conceptual issues and policy questions, *Journal of Urban Affairs*, 28(1), pp. 1–17.

Silver, D., Clark, T. N. & Grazuil, C. (2011a) Scenes, innovation, and urban development, in: D. E. Anderssson, Å. E. Andersson & C. Mellander (Eds) *Handbook of Creative Cities* (Cheltenham: Edward Elgar).

Silver, D., Clark, T. N. & Navarro Yanez, C. J. (2011b) Scenes: Social context in an age of contingency, in: T. N. Clark (Ed.) *The City as an Entertainment Machine,* revised ed. (Lanham, MD: Lexington Books).

Smit, A. J. (2011) The influence of district visual quality of location decisions of creative entrepreneurs, *Journal of the American Planning Association*, 77(2), pp. 167–184.

Storper, M. & Venables, A. J. (2002) Buzz: the economic force of the city. Paper presented at the DRUID Summer Conference on 'Industrial dynamics of the new and old Economy—who is embracing whom?', 6–8 June 2002, Copenhagen, Elsinore.

Trip, J. J. (2007) Assessing quality of place: A comparative analysis of Amsterdam and Rotterdam, *Journal of Urban Affairs*, 29(5), pp. 501–517.

Trip, J. J. & Romein, A. (2009) Beyond the hype: Creative city development in Rotterdam, *Journal of Urban Regeneration and Renewal*, 2(3), pp. 216–231.

UNDP/UNCTAD. (2010) *Creative Economy Report 2010.* United Nations.

Zimmerman, J. (2008) From brew town to cool town: Neoliberalism and the creative city development strategy in Milwaukee, *Cities*, 25(4), pp. 230–242.

Cultural Amenities: Large and Small, Mainstream and Niche—A Conceptual Framework for Cultural Planning in an Age of Austerity

ROBERT C. KLOOSTERMAN

Amsterdam Institute for Metropolitan and Development Studies (AISSR), University of Amsterdam (UvA), Amsterdam, The Netherlands

Abstract   *Cultural planning has been high on the agenda of many policy-makers. From an end in itself, it has been transformed into an instrument to regenerate neighbourhoods and even whole cities and as a means to boost the quality of place to attract high-skilled workers. With the current crunch on public spending, the question arises what will happen to cultural planning initiatives and what scope will remain for them. To explore what may happen to cultural planning in this age of austerity, we present a concise typology of cultural amenities based on two underlying, business model, dimensions. The first dimension concerns the supply side, namely the scale of provision of the cultural amenities. The second dimension, located on the demand side, is the market: orientation of the amenities: mainstream- or niche-oriented. Each type is associated with a specific location pattern, impact on the quality of place and funding configuration. We expect that the budget cuts will especially affect the small-scale, niche-oriented cultural amenities which are crucial for the quality of place. We also expect a trend towards further commercialization and commodification threatening the authenticity of the large-scale, niche-oriented cultural amenities.*

If public goods-public services, public spaces, public facilities-are devalued, diminished in the eyes of citizens and replaced by private services against cash, then we lose the sense that common interests and common needs ought to trump private preferences and individual advantage. (Judt, 2010, p. 129)

The question of balancing art and investment, aesthetics and consumerism will continue to plague politicians, artists and citizens all over the world. (Plaza, 2006, p. 464)

1. Introduction

On a very prominent location, right behind the Amsterdam Central Station on the other side of the IJ River, a new, gleaming iconic building can be seen. The elegantly designed building houses the Dutch national film museum or the "EYE Film Instituut" (http://www.eyefilm.nl/). The "EYE Film Instituut" (Figure 1), which opened its doors in April 2012, is responsible for taking care of some 37,000 old art films, about 700,000 pictures and numerous books, soundtracks and other film-related items. As such, the "EYE Film Instituut" serves as the national repository for films and film making in the Netherlands. It is funded by the Dutch state to the tune of 7,000,000 euros per year and by the Amsterdam municipality for a much more modest yearly amount of 50,000 euros. Whereas other cultural activities were hard hit by the recent fierce cuts in state subsidies, the "EYE Film Instituut" came out relatively unscathed. It is expected that the new museum will attract about 225,000 visitors every year (van Zwol, 2012). Public funding and the sale of regular tickets alone, however, will not pay all the bills. To balance the budget, the "EYE Film Instituut" has to include the organization of seminars, educational activities, parties and a whole array of commercial activities in its business model. The institute, moreover, also relies on the activities of about a 100 volunteers. The "EYE Film Instituut" is a vivid example of how cultural institutions in the Netherlands deal with the current funding climate and how they have to find a way out of the dilemma between, on the one hand, commercialization and, conceivably, dumbing down, and, on the other hand, maintaining a more elitist "l'art pour l'art" attitude and thus focusing on a potentially significantly smaller group of high-end visitors with a considerable stock of cultural capital.

Figure 1. The "Eye Film Instituut", Amsterdam.
Source: Picture by the author.

Those responsible for running cultural amenities are not the only ones faced with difficult choices. Local policy-makers and urban planners now also have to rethink their policies and strategies in this age of austerity that followed in the wake of the credit crisis of 2008. From the 1990s onwards, cultural planning—"the strategic and integrated planning and use of cultural resources in urban and community development" (Evans & Foord, 2008, p. 72)—has been high on the agendas of many urban policy-makers (van Aalst & Boogaarts, 2002; Miles & Paddison, 2005; Mommaas, 2004; Sacco & Crociata, 2012). Many of them were notably inspired by the successes of the "Guggenheim Museum" in Bilbao where a flagship building kick started an urban regeneration process and by the "Tate Modern" in London which put the south bank of the Thames on the map. Cultural planning became even more important after Bianchini and Landry (1995), Landry (2008) and Florida (2002) stressed the importance of cultural amenities which may increase the "quality of place", draw high-skilled workers and subsequently boost the urban economy (Evans & Foord, 2008, p. 71). Even in an age dominated by a neoliberal agenda and a concomitant retreat of the state from the field of urban planning (Peck, 2012), "[t]he concept of the creative city ... has become a powerful talisman for urban planners. Cultural policy has much to contribute towards re-vitalising depressed urban areas, improving liveability, and stimulating urban and regional economic growth" (Throsby, 2010, p. 29).

Cultural amenities—the set of institutions (public and private) which enable the "local" consumption or provision of services with a high semiotic or aesthetic value such as museums, galleries, zoos, theatres, festivals and sport venues—are then important for contemporary urban economies (Clark & Kahn, 1988, p. 363; Evans, 2009, p. 1008; Scott, 2004, p. 462). What, then, are the consequences of the current financial constraints for cultural planning? As Evans and Foord (2008, p. 65) have stated, a "...growing demand for an informed framework for planning arts and cultural facilities has emanated from both local and regional government as well as cultural sectors". This need has become, arguably, more pressing with the tightening of budgets in the public sector after 2008 which has significantly reduced the scope for cultural planning initiatives.

The aim of this contribution then is two-fold. First, we present a simple typology of cultural amenities in an attempt to address this "growing demand". Second, we explore the consequences for cultural planning and cultural amenities in an age of austerity on the basis of this typology: which kinds of cultural amenities are important from the viewpoint of cultural planning and should be given priority? The typology presented below departs from two crucial business model characteristics: scale of operation on the supply side and market segments on the demand side. The thus constructed ideal-typical categories can subsequently be related to various other characteristics of cultural amenities as location patterns, differences in potential local impact with respect to quality of place and different abilities in generating their own incomes thereby suggesting diverse divisions of labour between the public and the private sectors. These differences are highly relevant for urban planners and for those running cultural amenities.

We first put post-war cultural planning briefly into perspective and highlight different phases from state-led paternalism and cultural amenities as goal in itself, to culture as a means to economic ends and, more recently, to the rollback of the public sector, nationally and locally, with respect to supporting cultural amenities (Section 2). We then present the four types of our conceptual framework of cultural amenities (Section 3). After that, we examine the implications of our framework to address the question what the consequences are of the current phase of a retreat of state funding for cultural amenities and cultural planning initiatives and try to identify a range of options for public and/or private funding given the nature of the cultural amenity and its urban context. In the conclusions, we will wrap up our findings and put them in a broader perspective (Section 4).

2. Cultural Planning as a Strategic Local Economic Policy

Governments—local and national—have been interfering in culture for ages. Pharaohs, kings, emperors, dictators, and democratically elected governments have used culture to impress people. In that sense, cultural planning is nothing new under the sun. A closer look at means and ends of cultural planning, however, does reveal salient differences and shifts over time. Cultural planning, to use a rather recent definition, is "... the strategic and integrated planning and use of cultural resources in urban and community development" (Mercer, 2006, p. 6). Cultural planning has not only differed over time, but also across places. The differences in the institutional embeddedness of cultural planning in the US and many European countries are obvious with much larger role for the state in the latter than in the former. There are also significant differences between European countries; the French state-centred approach is quite different from the more private-sector approach in the UK (Sassoon, 2006). Below, we offer a helicopter view which neglects these differences. Instead, we focus on the more general changes in the wider configuration of cultural planning after the Second World War. Point of departure for our brief sketch of the key developments is the overview of urban planning strategies offered by Evans and Foord (2008, p. 71) and we distinguish four phases.

The first phase is that of the post-war era lasting until about the mid-1970s. Many governments in the years after the Second World War came to see culture as in the same vein as education and healthcare which had to be distributed across the population, socially and spatially (i.e. they viewed culture as a merit good: to make it available to everyone because they believed it to be important; Towse, 2010, p. 34). The historian Judt (2010, p. 53) has called this cultural policy a "uniquely successful blend of social innovation and cultural conservatism" and Keynes in his view "... exemplifies the point. A man of impeccably elitist tastes and upbringing ... he nonetheless grasped the importance of bringing first-class art, performances and writing to the broadest possible audience to overcome its paralyzing divisions". This quite paternalistic approach was part of a more comprehensive modernist attempt to make society more civilized. Making culture more accessible in all kinds of ways to the entire population was seen as an important public task. Theatres, museums, regional orchestras, performing arts; all should contribute to the distribution of the higher arts among large segments of population who, apparently, were not fully aware of its importance. At that time there was not the slightest doubt about the definition or domain, or more precisely what belonged to higher culture. The cultural hierarchy was still in place and the elite seemed to know which culture should be distributed. Civic cultural centres and neighbourhood facilities, first as part of larger master plans and later on more bottom-up, were much in vogue (Evans & Foord, 2008, p. 71). The principle of state funding for cultural amenities was not very much contested (Hobsbawm, 1996, p. 508).

In the phase thereafter, roughly from the mid-1970s to the mid-1990s, a rather radical change occurred. Instead of a goal in itself, "...culture became more of an instrument in the entrepreneurial strategies of local governments and business alliances" (Zukin, 1995, p. 12). This shift was related to deep-rooted changes whereby modernist cities gave way to "postmodernist" cities (Harvey, 1989) and which occurred against the backdrop of the unravelling of the Keynesian welfare state and the emergence of neoliberalism. Many cities were in crisis with both a decline of population and a decline of employment. Cities became fierce competitors for the attraction of firms and tourists to their territory (Gospodini, 2002). In many European cities, culture came to be seen as an instrument to strengthen the local economy, brand the city, and as a crucial plank in the strategies for inter-urban competition that was heating up at that time (Le Galès, 2002, p. 221).

These changes more or less coincided with the erosion of the distinction between what was once considered as high and low culture: "The very distinction between 'high' and 'low' culture was itself looking increasingly shaky, a product of an earlier era of elite intellectual self-confidence and benevolent moral superiority" (Hobsbawm, 1996, p. 513; Mazower, 1998, p. 354). As high and low culture had been clearly separated by ways of funding—high culture mainly funded by the state and low culture provided by the private sector—this seemingly clear separation could not be made anymore. A new understanding of cultural planning thus emerged. Instead of spreading the high culture among large segments of the population, the focus was now on cultural amenities such as flagship projects (museums and opera houses) which would contribute to the image of the city and, hence, to attract tourists and, preferably, firms as well (Evans & Foord, 2008, p. 71). Cultural planning became subsequently a strategic instrument in local economic policies.

The third phase, from the mid-1990s to 2010, constituted no fundamental break with the second phase apart from its emphasis in cultural planning on boosting the quality of place to attract or retain high-skilled workers or to use Florida's (2002) vocabulary members of the creative class. In this phase, the relationship between these particular segments of the urban labour force was highlighted and the cultural amenities were seen as a more integral part of the production milieu. Cultural amenities together with shops, cafés and restaurants created not only a particular atmosphere of urbanity, diversity, excitement and tolerance for innovative or creative workers (van Aalst, 1997; Florida, 2002; Helbrecht & Dirksmeier, 2012), but also provided the setting for the exchange of (tacit) knowledge in meaningful face-to-face encounters: "Art museums, boutiques, restaurants, and other specialized sites of consumption create a social space for the exchange on which business thrive" (Zukin, 1995, p. 13). In the third phase, then, there was more awareness of the linkages between street-level small-scale amenities, shops, restaurants and cafés and small firms in the cultural industries.

The causality of the relationship between cultural amenities and high-skilled workers is highly contested. Repudiating Florida's more voluntaristic view, Scott (2008) has argued that the production system comes first and that cultural amenities are very much part of subsequent emergent effects. The relationship is, arguably, more interactive than either Florida or Scott has proposed. The production system is indeed of great importance, without it no creative workers. However, if everything else in terms of the production system can be considered equal, differences in cultural endowments and quality of place could give one city the edge over another (Kloosterman & Trip, 2011). The causal relationship with the so-called creative class may be contested, but the empirical evidence for a relationship between culture and quality is quite convincing. The strong economic position of Amsterdam in the Netherlands and its relatively large share of high-skilled workers are clearly correlated with its rich palette of cultural amenities (de Groot *et al.*, 2010, p. 73). Evans (2009, p. 1009) observes that "[t]he association between quality of life, amenities and inward investment/firm relocation was also established in these early studies". With the creative city perspective, the focus of urban planning moved not just away from an emphasis on physical infrastructure (e.g. highways, airports), but also from flagship projects to the rich fabric of the local cultural infrastructure including both small-scale and alternative cultural amenities. Quality of place and especially culture has become, consequently, more and more an issue in urban planning (Trip, 2007).

After 2010, however, the scope for this kind of cultural planning was drastically reduced as a new phase was entered, a phase of austerity. Investing in cultural amenities still makes sense as issues of quality of place and its relationship with high-skilled workers are still relevant. Moreover, EU policy rules preclude many other growth-stimulating policies such as direct subsidies to firms. The effects of investments in culture will not, in addition, easily leak out of the region. In addition, one can point to the fact that, in contrast to investments in the transportation infrastructure or in prestigious office parks, spending money on culture can often count on much sympathy in local communities (and, therefore, among voters). What, then, is the scope for cultural planning in age of austerity?

3. Cultural Amenities: Small and Large, Niche and Mainstream

Cultural amenities encompass a large variety of activities. They comprise, as said, museums, theatres, galleries, zoos, sports venues, festivals and other activities that enable local consumption of assorted cultural services. We could look at the impact on quality of place for each type separately, but to explore the potential impact of cultural amenities and their contribution to the quality of place, we need a more analytical approach that covers, in principle, all these types. To make that more general point, we take a closer look at museums in Amsterdam. A museum may display the grandeur of the "EYE Film Instituut" catering to the film cognoscenti. Or it may be the large "Madame Tussauds Amsterdam" where "you can come face to face with your favourite celebrities!" at the Dam square in the heart of the city apparently appealing to those who have at least some basic knowledge of the tabloids. Or, it may be the very small "Max Euwe Centrum", dedicated to the only chess world champion the Netherlands has ever had (1935), attracting mostly die-hard chess fans. The impact of these three specimens on the quality of place in Amsterdam and their contribution to attracting high-skilled workers will, obviously, diverge significantly. These differences are obviously not limited to museums alone, other cultural amenities may also display considerable variability in scale of operation and in the type of visitors they predominantly attract. The characteristics of the business models of various cultural amenities, hence, may differ considerably.

Our point of departure is that by distinguishing two, in principle, independent axes of the business model a more strategic differentiation of cultural amenities can be obtained. The first one is a "supply-side" characteristic and concerns the scale of operation, while the second one covers the "demand side" by focusing on the potential market of the amenity: more mainstream- or more niche-oriented. We, thus, unpack the container concept of cultural amenities from an economic viewpoint into four different types. Each of these types has its own socio-economic and spatial logic and, accordingly, its own specific potential impact on the local economy in terms of quality of place. In addition, each ideal-typical cultural amenity can also be related to different configurations of potential sources of income or funding. Before zooming in on the four types, we briefly explicate the two dimensions.

Cultural amenities, as the examples of the museums made clear, can differ considerably in terms of the scale of operation. This can be traced to the indivisibility of some cultural amenities. There are, for instance, clearly limits to the extent one can economize on the size of a symphony orchestra (minimum size about 100 players) or a chamber orchestra with (about 40 players) (Towse, 2010, p. 227). The large collection of films and film-related items of the "EYE Film Instituut" also requires a certain minimum scale in terms of housing, maintenance, arranging and labelling. In addition to these substantial indivisibilities, there are also economies of scale associated with the provision of cultural amenities (Clark & Kahn, 1988, p. 363). If fixed costs are high, as, for example, in the case of performing an opera, economies of scale arise and provision at large scale will be more efficient. Not all cultural amenities display substantial indivisibilities or are prone to econ-omies of scale. A chess museum or a small gallery specializing in seventeenth-century Dutch Delftware and polychrome animals and figures are much more likely to suffer from diseconomies of scale as enlargement will dilute the focus and exclusivity and may, in the case of the gallery, harm personal relations with the customers.

The other dimension concerns the demand side or the composition of the clientele. Tra-ditionally, the main distinction has been between "the arts" or "high culture" on the one side and "popular culture" on the other side. However, "[t]he rigid distinction between high art and popular culture that permeated arts policy in earlier times and that inevitably identified it with the upper echelons of society has been gradually relaxed" (Throsby, 2010, p. 83). With "the stabilisation or deinstitutionalisation of former taste hierarchies", this dichotomy has thus become increasingly blurred (Mommaas, 2004, p. 517, see also Sacco & Crociata, 2012, p. 4). This blurring, however, has not done away with the need to classify culture. Consequently, new dimensions have been suggested to distinguish between different cultural products and consumption practices as, for example, commer-cial/non-commercial, traditional/avant-garde, mass/specialized and majority/minority (Throsby, 2010, p. 2). The underlying dimension of these dichotomies here seems to be the distinction between, in principle, larger, mainstream audiences and smaller niche audi-ences. Cultural amenities may be targeting more mainstream audience, as in the case of the "Madame Tussauds Amsterdam" or a zoo, or cultural amenities may be aimed at specific niches requiring, in principle, a particular type of cultural capital, in the sense of a particu-lar know-how which allows a person to appreciate and enjoy the cultural service on offer. These niches may be high-brow, as in the case of the chess museum, or they may be cater-ing to specific subculture as in the case of the "Hash, Marijuana and Hemp Museum" in Amsterdam without being necessarily high-brow. Admittedly, this distinction is not iron-clad, as many forms of culture are layered in their semiotic content and can be enjoyed on different levels of interpretation and, hence, by a variety of people in terms of cultural capital. Paintings by Jan Steen or David Hockney, music by Bach or Elbow can be enjoyed in quite different ways and, hence, can appeal to broader segments of the audience. Things get even more complicated if we take into account Bourdieu's (1984) view of continuously shifting boundaries as cultural elites are in constant search of cultural goods or services that will set them apart from the masses. We will come back to these dynamics below.

Notwithstanding these complications, it still makes sense to start with relatively simple typology of cultural amenities and then work out the consequences. Combining the two dimensions generates four types of cultural amenities (Table 1). The dimensions refer, of course, to continuums, but for the sake of simplicity they have been reduced to dichotomies. Below, we will discuss each type more in detail and look at the locational characteristics, the catchment area, the potential contribution to quality of place and attraction of visitors, residents, workers and firms. In addition, we will also look at the "business model" to assess if and how this type of amenities should be part of a state-led form of cultural planning.

The first type of cultural amenities is small scale and caters to niche markets which require a certain knowledge or cultural capital to grasp and appreciate what is on offer. No large collections to maintain, no high fixed costs, no need for large premises and no scores of people to employ make provision on a relatively small scale feasible. Small museums or galleries with small, specialized collections would, for example, fall under this heading. Specialized does not necessarily entail what is more traditionally labelled as "high culture"; a "gothic" event would also be included in this category, although there obviously is a strong overlap.

This kind of cultural amenities can, at first glance, be located anywhere. No great demands regarding infrastructure or premises should make small-scale cultural amenities relatively footloose. Indeed, we find galleries specializing in sculptural art located outside cities. This is, however, only a small part of the story. Being dependent on niches, these cultural amenities can clearly benefit from agglomeration economies. Large concentrations of people will contribute to the creation of sufficiently large diverse customer bases to maintain niche-directed cultural amenities and may set in motion a process of deepening of the division of labour an specialization among these amenities leading to ever more refined niches with each individual amenity benefiting from the proximity of the others—a process that can be observed for restaurants as well (Glaeser, 2011, p. 123; Steel, 2008, p. 150). Put it more bluntly: the larger the city, the higher the number of specialized cultural amenities and the higher the level of specialization (Poon & Lai, 2008). Given their small scale, these amenities can locate and cluster in mixed-use neighbourhoods, thereby contributing to a complex urban fabric interwoven with small shops, restaurants and meaningful public spaces and, hence, giving cities a special flavour (Santagata, 2002; Scott, 2000, 2004). Often, there are many hybrids straddling more mundane forms of consumption with more explicit cultural aims as in the case, for instance, of restaurants which also serve as art galleries or venues for performances.

If present in sufficient numbers, small-scale, niche-oriented cultural amenities can contribute to the quality of place not just in the neighbourhood where they are located, but even for the city as whole. As stated by Poon and Lai (2008, p. 2276): "cities with

Table 1. An economic typology of cultural amenities

	Scale of provision	Small	Large
Type of audience	Niche	Art galleries	"EYE Film Instituut"
		Modern dance performance	Van Gogh Museum
	Mainstream	Erotic Museum	"Madame Tussauds Amsterdam"
		Popular music performance	Heineken Experience Zoo

diverse amenities have become more attractive places to live in". High-skilled, "cognitive-cultural" workers (Scott, 2008), members of the so-called creative class (Florida, 2002) as workers, (potential) residents and visitors, then, will be attracted by a rich and diversified urban milieu and by the vibrancy in the city (partly) created by these small-scale amenities. One could even argue that these amenities are part and parcel of a wider creative field encompassing "... an atmosphere and a common set of resources, creating a platform for creative and innovative activities" (Scott, 1999, p. 809). Small-scale, specialized cultural amenities constitute, accordingly, important ingredients in determining the quality of place and, hence, the attractiveness of cities. They are also important for the support and the renewal of cultural industries as they serve as incubators and places to meet. With regard to their strategic potential, should small-scale, niche-oriented cultural amenities be part of cultural planning and, if so, how should they be fostered?

Given their small size and their specialist orientation, planning of these kinds of cultural amenities does not make much sense. Local governments lack the specialist, fine-grained knowledge to interfere directly and should leave this to local cultural entrepreneurs who do have that knowledge and who are willing to take the risks and create or fill a specific niche. It does make sense for local governments, however, to create the conditions under which these cultural amenities can thrive. Small-scale cultural amenities are often dependent on cheap office spaces. Jane Jacobs once sang the praise of "aged buildings". According to her:

> Well-subsidized opera and art museums often go into new buildings. But the unfor-malized feeders of the arts—studios, galleries, stores for musical instruments and art supplies, backrooms where the low earning power of a seat and table can absorb une-conomic discussions—these go into old buildings ... Old ideas can sometimes use new buildings. New ideas must come from old buildings. (Jacobs, 1961, p. 181)

For cities with a sizeable and well-preserved historical core, old buildings do not necess-arily provide cheap spaces, sometimes quite on the contrary, but the general point about the dependence of small-scale cultural amenities on low-cost spaces remains relevant. It is not just the costs of these spaces but also their openness towards users: "Cheap spaces that can be innovatively adapted to reduce financial risk and encourage exper-iment" (Brandellero & Kloosterman, 2010; Landry, 2008, p. 123). Consequently, to foster small-scale, niche-oriented cultural amenities that contribute to quality of place, ensuring the provision of cheap spaces which can be used in many, even unexpected ways will be an important plank in cultural planning strategies. In the wake of deindustria-lization, such spaces were in many cities abundant as factories closed, shunting yards were abandoned and docklands were vacated. Old industrial buildings located in the city centre or quite close to them became available for other uses after the 1980s. Existing forms of built environment allowed very different functions as lofts, factories and warehouses were converted to apartments or to incubators housing cultural or creative activities. Small-scale cultural amenities together with artists often spearheaded gentrification pro-cesses (Ley, 2003). However, during the boom years after 2000, another lesson of Jane Jacobs, that of the self-destruction of diversity, was also brought home, when cities seemed to run out of such cheap spaces. In, for example, New York this threatened to undermine the very cultural fabric that had helped the city to back on its track in the 1980s (Currid, 2007).

Strategic cultural planning aimed at making cities attractive for residents, workers and visitors by promoting small-scale, niche-oriented cultural amenities should make sure that within the city as a whole cheap spaces are available for various, often unforeseen uses. On the level of individual neighbourhoods, market pressures may initiate processes of self-destruction and cheap spaces may be pushed out, on the level of the city as a whole; however, cheap spaces should still be available either because of local slack, slow turnover or even conscious intervention by the city itself. Ensuring cheap spaces is partly a matter of zoning, as new, different uses are sometimes blocked by regulations. The current wave of shop closures in city centres due to the rise of internet shopping is now creating a surge of empty spaces which are, in principle, suitable for other uses including cultural amenities. Zoning plans should be adjusted to allow these new uses. In addition, local governments may aim at improving the infrastructure by boosting entrepreneurship among those who do have the cultural capital to cater to niches. Instead of pursuing a career as an artist, it might make more sense for many would-be painters, musicians and actors to start establish a cultural amenity and enrich the infrastructure of a city.

Small-scale cultural amenities which do not put high demand on its users in terms of cultural capital and, accordingly, cater to local mainstream audiences are a rather difficult category. In smaller cities and towns, they can survive, being, to some extent, protected against competition by distance. A local museum, in which the underlying common element of the collection is the link with a particular place (from local archaeological findings and stuffed birds to the work of local artists), can serve as an example of a cultural amenity which is accessible to large groups and provided on a relatively small scale. Local needs are met by this type and people are usually not willing to travel long distances for these amenities and people of the creative class cannot be expected to be very interested. Local identity and social cohesion can be boosted by these amenities by mobilizing and integrating civilians in the local community. Dedicated local actors, both private and public, can take the initiative to establish and maintain these kinds of amenities not just by funding the premises but by covering (part) of the personnel costs as well.

In larger cities, small-scale, mainstream-oriented cultural amenities are not protected by distance. They have to compete either with specialized amenities or with much larger ones benefiting from economies of scale. The expected dynamics for these kinds of amenities in more urban contexts is then either to specialize and move away from the mainstream or scale up and become larger. Small-scale, mainstream-oriented cultural amenities should, therefore, only be included in cultural planning policies in small settings and not in larger cities.

The third type of cultural amenities comprises the kind of iconic projects so beloved by many urban policy-makers: large scale and catering to the demanding taste of connoisseurs. Much cultural planning is focused on these striking cultural amenities which contribute to the identity and international image of a place (place-making). They do not just require large buildings, but they are typically housed in high-profile flagship buildings. New buildings designed by so-called starchitects, as in the case of the "EYE Film Instituut" in Amsterdam, the Disney Hall in Los Angeles and the inescapable "Guggenheim Museum" in Bilbao; or converted industrial buildings as the "Tate Modern" housed in former power station in London and the "Musée d'Orsay" in Paris; or recently renovated buildings as the "Stedelijk Museum" in Amsterdam are striking examples of these kinds of amenities (Kloosterman, 2010).

Apart from reasons of intrinsic cultural value, as in the case of preserving the Dutch film heritage in the "EYE Film Instituut", these museums also make economic sense. According to Frey and Meiers,

> There are two types of demand for museums. The first is the private demand exerted by the visitors. These may be persons interested in the exhibitions as a leisure or as part of their profession as an art dealer ... The second type of demand comes from persons and organizations benefiting from a museum, but not expressing their demand at the cashier's office. This social demand is based on external effects and/or the effects of art organizations on other economic activities. (Quoted in Plaza, 2006, p. 460)

The private demand can be quite substantial, notwithstanding the relatively high threshold in terms of cultural capital is relatively high, as these amenities can rely on large catchment areas of national and often even international scope. Admirers of these palaces of "high" culture are willing to travel long distances to enjoy this type of cultural amenities generating thus sufficient critical mass. These amenities—not just the buildings but the whole package together with the collections or the performances—should, then, be so unique that people are indeed willing to travel far to enjoy it. As visitors do not bother to travel hundreds or even thousands of kilometres just for visiting the museum, these visits are regularly part of city trips that may take several days.

To meet the demands of these visitors, this type of cultural amenity characteristically definitely implies an urban setting. A good transport infrastructure (including a nearby airport) and provisions like hotels and restaurants are needed. For a more permanent competitive position, a differentiated supply of a whole range of provisions, notably comprising small-scale, niche-oriented cultural amenities but also (specialized) shops, is essential. Agglomeration economies, consequently, kick in and larger cities with a broader range of amenities and good accessibility can strengthen their position as (international) travel destination by initiating and supporting large-scale, niche-oriented flagship cultural amenities.

The social demand for a museum is based on the external or spill-over effects. As the potential indirect effects of a rise in the quality of place resulting in more (high-skilled) workers wanting to work and live there and more tourists wanting to visit the city (e.g. expenditures on hotels, restaurants and shops) are hard to internalize by the cultural amenities themselves, we are confronted with a classic case of a market failure (Throsby, 2010, pp. 35–37). Local governments often step in these museums, at least in many European settings, and tend to rely on forms of public funding (Plaza, 2006). In the case of the "Tate Modern", the external benefits are even deemed so substantial that the museum does not even bother to levy an entrance fee. The success of the "Tate Modern" shows that strategic investment in a large-scale, in principle niche-oriented (in this case high-end modern art) cultural amenity can make good economic sense, not just to put a city as Bilbao on the map of international cultural tourism but also to strengthen a city's already strong position as a tourist destination. The "Tate Modern" also makes clear that the overall urban context is crucial in maximizing the positive external effects. To reap these benefits on a more permanent basis, hence, a city needs as said a thick infrastructure of small-scale cultural amenities, shops, restaurants and hotels. The emergence of this kind of infrastructure is not given, but depends on the presence of entrepreneurs who have the cultural capital to cater to the (changing) needs of the workers, residents, visitors and tourists (Plaza, 2006, p. 464). Larger cities with a strong historical pedigree in culture and arts—such as Amsterdam, London and Paris, tend to reproduce this infrastructure over longer periods (Deinema & Kloosterman, forthcoming). According to Throsby (2010, p. 134), "Moreover, the flow of services from this stock of tangible and intangible cultural capital, which in these cities has been generated continuously over centuries, is of a particular self-reinforcing kind where, in short, art creates art."

Given the external effects, large-scale niche-oriented cultural amenities can be part of successful cultural planning strategies. Governments may use the establishment of new museum or the renovation of existing museums to boost the quality of place and foot a significant part of the bill. The impact of such an endeavour in the long run is, however, dependent not just on the design of the museum itself, but also on the wider social and economic context. According to Miles and Paddison (2005, p. 837), "The single most dangerous aspect of cultural investment is that it simply does not sit comfortably in the context for which it is intended."

The fourth type of cultural amenities concerns large-scale "mainstream" cultural amenities, or in other words "mass culture" which can be enjoyed without much specific knowledge of the content presented. Examples of this kind of these large, low threshold cultural amenities are zoos, theme parks and large venues for musicals and pop-concerts. Because of the size necessary to accommodate large audiences, fixed costs tend to be high and, consequently, economies of scale are prevalent. These amenities usually take up large spaces and good accessibility by car, public transport or both is crucial for the functioning of these amenities.

In contrast to the niche-oriented cultural amenities, large-scale mainstream amenities typically tend to be located not in the centres of larger cities, but, at least in the West-European context, on the outskirts near highways or, as in the case of space-consuming theme parks, even at some distance of cities. This relative spatial isolation makes it, on the one hand, much easier to internalize the spill-over effects of spending by visitors, but, on the other hand, diminishes the impact on the city itself. For this reason, and because mainstream offerings do not add much distinction, the quality of place, then, is not much affected by these amenities.

Aiming at mainstream audiences, concentrated at more or less confined areas, moreover, can be quite profitable. There, then, is no market failure and no compelling reason why government should step in to help supplying these amenities apart from planning the locations and providing (part of) the infrastructure. The responsibility should lie with the private sector which should be able to organize and manage these kinds of amenities commercially.

4. Conclusions

Culture, whatever its exact form, plays an increasing role in our lives. The on-going process of economization of culture refers not just to culture being increasingly a product which can be bought and sold in markets and which is part of the expanding set of cultural industries, but also to culture as an instrument in (local) economic policies. As Dowling (1997, p. 30) observed already more than a decade ago: "Putting culture on the urban planning agenda ... has been crucial in illustrating the centrality of culture in everyday life." In the post-war years, cultural planning in many West-European countries was aimed at distributing what was seen as high culture both socially, to the lower echelons of the population, and spatially, to other places than just the larger cities or the cultural capital (Sassoon, 2006).

In the 1970s and 1980s, however, culture began to transform from a goal in itself to a means to an end, at first defined in both social (community development) and economic terms (Evans & Foord, 2008, p. 71). Gradually, the goal and even the "raison d'être" of culture became more and more conceived in sheer economic terms. Regeneration of neighbourhoods, putting a city on the (tourist) map, and, more recently, attracting high-skilled workers and, more fanciful, luring members of the creative class by offering a high quality of place with a wide range of cultural amenities have become the key aims to which investing in culture is seen to contribute. This shift to culture as a means to an end occurred while neoliberal urban policies stressing the importance of the private sector and striving to roll back the public sector became prevalent in many European cities. The linkages between cultural planning, the creative city and the neoliberal surge are complex and manifold (Peck, 2012) and may differ according to place and time. We can, however, identify a few common elements: the emphasis on competition in general and between cities more in particular; betting on winners and a drift away from programmes aimed at social and spatial redistribution and instead aiming at those social groups and cities which are already relatively well-endowed; the importance of an entrepreneurial approach and a move away from high culture as the end-all-and-be-all of cultural planning. This, then, has amounted to a more or less reversal of the early post-war policies which according to Tony Judt could be characterized as a combination of "social innovation and cultural conservatism". More recent policies are better seen as combination of social conservatism and cultural innovation in order to increase the quality of place and attract high-skilled workers.

After the outbreak of the credit crisis, budget cuts—affecting in particular spending on cultural amenities—on the national and the local level have been fierce in among other countries, the UK, Spain, Ireland and the Netherlands. Cultural planning has obviously entered a new phase. Above, we have tried to offer a transparent conceptual framework intended as a starting point for analyzing which role (local) governments can and should play when considering strategic cultural planning aimed at strengthening a city's economic base. The underlying dimensions of this framework are, respectively, the scale of the provision and the market orientation. Four categories of ideal-typical cultural amenities are then identified: small scale/niche-oriented, large scale/niche-oriented, small scale/mainstream-oriented and large scale/mainstream-oriented. Social reality is, evidently, less neat and boundaries are much more blurred and vague and the framework resembles more a field with four poles in which more intermediary positions and complex combinations are possible. The "Tate Modern", for instance, notwithstanding its high-brow modern art collection, has, arguably, become part of the experience economy (Pine & Gilmore, 1998) and able to attract a very broad range of visitors who are certainly not all connoisseurs of modern art (Dalley, 2010). We can use this framework, however, to present a broad and hypothetical analysis of the effects of the impending cuts.

What are the tentative implications that can be gleaned from the above analysis? Small-scale/mainstream-oriented amenities, it seems, will have a hard time to survive in urban environments. They either have to enlarge the scale to survive price competition or move to a niche to avoid that kind of competition and—apart from small towns and villages—should not be the object of cultural planning. The same can be said with respect to large-scale/mainstream-oriented amenities, which can be left to themselves in terms of funding as they are able to generate their own incomes and generally do not generate much positive externalities in terms of boosting the quality of place and attracting higher skilled workers. Large-scale/niche-oriented cultural amenities as the "EYE Film Instituut" in Amsterdam typically have more difficulties in generating their own income as they are dependent on larger catchment areas than mainstream-oriented amenities, but they may, on the basis of their more or less unique offerings, contribute to the quality of place. They are already very much part of cultural planning strategies and given their potential for positive externalities this makes sense even in age of austerity. The impact of these large-scale projects, we expect, will also be contingent on to what extent they are part of a larger fabric of small-scale, niche-oriented amenities which gives cities not just their particular flavour but which may also cater to a variety of workers, residents and visitors. Cultural planning, then, should aim at helping to create the spaces—even literally—for these bottom-up amenities in terms of zoning and, arguably, rent control. Glaeser (2011, pp. 66–67) has also pointed to the danger of neglecting this infrastructure of smaller amenities which are crucial in making places attractive to high-skilled workers:

> Museums and transportation and the arts do have an important role in place-making. Yet planners must be realistic and expect moderate successes not blockbusters. Realism pushes towards small, sensible projects, not betting a city's future on a vast, expensive roll of the dice. The real payoff of these investments in amenities lies not in tourism but in attracting the skilled residents who can really make a city rebound.

On the basis of the model, we might also expect shifts within the typology driven by the cuts in public funding. A push towards opening up to mainstream audiences implying a further commodification, commercialization and shift to consumerism seems in the offing. To go back to the "EYE Film Instituut" once more, this large-scale, niche-oriented cultural amenity lets its restaurant with its great views on the river IJ and the Amsterdam skyline on the other side, for weddings and other parties to generate sufficient income. More and more, we will see that cash-strapped, niche-oriented cultural amenities, both large and small, become more entrepreneurial and offer services to open up to more mainstream audiences in order to safeguard its more niche-oriented aims or become more mainstream-oriented altogether. There is nothing inherently wrong in becoming more entrepreneurial and more market-oriented, but the looming danger is one of dumbing down and selling out. This might materialize in a process of Disneyfication and undermine the authenticity and distinctiveness of the cultural amenities and, hence, erode the potentially positive contribution to the quality of place.

Apart from this critique on the possible failure of this instrumental aspect of cultural planning, one can deplore the loss of a political will to uphold a public domain of cultural provision on the basis of the intrinsic value of cultural amenities and their expression of communal pride fenced off from more mundane market considerations (Judt, 2010, p. 129). The loss of conviviality in contemporary cities (Scott, 2011) is partly a consequence of the subjugation of public spaces to market forces. Even in age of (relative) austerity, cultural planning should be aimed at guarding at least some of the "public" character of cultural amenities. Much poorer societies in the 1950s were able to do that, why should not we be able to make a small shift from private consumption to investments in the quality of public urban spaces not just for economic purposes, but also as a goal in itself?

Acknowledgements

This article is partly based on an earlier version which was co-written by Merijn van der Werff. I would like to thank Merijn for his contribution to the first version. I would also like to thank Len de Klerk, Allan Watson and two anonymous reviewers for their comments.

References

van Aalst, I. (1997) *Cultuur in de stad: Over de rol van culturele voorzieningen in de ontwikkeling van stadscentra* (Utrecht: Van Arkel).

van Aalst, I. & Boogaarts, I. (2002) From museum to mass entertainment: The evolution of the role of museums in cities, *European Urban & Regional Studies*, 9(3), pp. 195–210.

Bianchini, F. & Landry, C. (1995) *The Creative City* (London: Demos).

Bourdieu, P. (1984) *Distinction: A Social Critique of the Judgement of Taste* (London: Routledge).

Brandellero, A. M. C. & Kloosterman, R. C. (2010) Keeping the market at bay: Exploring the loci of innovation in the cultural industries, *Creative Industries Journal*, 3(1-2), pp. 61–77.

Clark, D. E. & Kahn, J. (1988) The social benefits of urban cultural amenities, *Journal of Regional Science*, 28(3), pp. 363–377.

Currid, E. (2007) *The Warhol Economy: How Fashion, Art and Music Drive New York City* (Princeton, NJ: Princeton University Press).

Dalley, J. (2010) Tate at 10, *Financial Times, Life & Arts*, May 1-2, pp. 10–11.

Deinema, M. N. & Kloosterman, R. C. (forthcoming) Historical trajectories and urban cultural economies in the Randstad megacity region. Cultural industries in Dutch cities since 1900, in: J. Klaesson, B. Johansson, C. Karlsson & R. Stough (Eds) *Metropolitan Regions: Preconditions and Strategies for Growth and Development in the Global Economy* (Berlin: Springer Verlag).

Dowling, R. (1997) Planning for culture in urban Australia, *Australian Geographical Studies*, 35(1), pp. 23–31.

Evans, G. (2009) Creative cities and urban policy, *Urban Studies*, 46(5/6), pp. 1003–1040.

Evans, G. & Foord, J. (2008) Cultural mapping and sustainable communities: Planning for the arts revisited, *Cultural Trends*, 17(2), pp. 65–96.

Florida, R. (2002) *The Rise of the Creative Class and How It's Transforming Work, Leisure, Community and Everyday Life* (New York: Basic Books).

Glaeser, E. (2011) *The Triumph of the City* (London: MacMillan).

Gospodini, A. (2002) European cities in competition and the new "uses" of urban design, *Journal of Urban Design*, 7(1), pp. 59–73.

de Groot, H., Marlet, G., Teulings, C. & Vermeulen, W. (2010) *Stad en land* (Den Haag: Centraal Planbureau).

Harvey, D. (1989) *The Condition of Postmodernity* (Cambridge: Blackwell).

Helbrecht, I. & Dirksmeier, P. (2012) New downtowns: A new form of centrality and urbanity in world society, in: I. Helbrecht & P. Dirksmeier (Eds) *New Urbanism; Life, Work, and Space in the New Downtown*, pp. 1–21 (Farnham: Ashgate).

Hobsbawm, E. (1996) *The Age of Extremes; A History of the World, 1914–1991* (New York: Vintage Books).

Jacobs, J. (1961) *The Death and Life of Great American Cities* (New York: Vintage Books).

Judt, T. (2010) *Ill Fares the Land; a Treatise On Our Present Discontents* (London: Allen Lane).

Kloosterman, R. C. (2010) Die Lehre aus Amsterdam: Neue Urbanität in der alten Stadt, *Geographische Zeitschrift*, 97(2+3), pp. 113–129.

Kloosterman, R. C. & Trip, J. J. (2011) Planning for quality? Assessing the role of quality of place in current Dutch planning practice, *Journal of Urban Design*, 16(4), pp. 455–470.

Landry, C. (2008) *The Creative City. A Toolkit for Urban Innovations* (London: Earthscan).

Le Galès, P. (2002) *European Cities; Social Conflicts and Governance* (Oxford: Oxford University Press).

Ley, D. (2003) Artists, aestheticisation and the field of gentrification, *Urban Studies*, 40(12), pp. 2527–2544.

Mazower, M. (1998) *Dark Continent; Europe's Twentieth Century* (New York: Vintage Books).

Mercer, C. (2006) Cultural planning for urban development and creative cities. Available at http://www.culturalplanning-oresund.net/PDF_activities/maj06/Shanghai_cultural_planning_paper.pdf (accessed 8 April 2012).

Miles, S. & Paddison, R. (2005) Introduction: The rise and rise of culture-led urban regeneration, *Urban Studies*, 42(5/6), pp. 833–839.

Mommaas, H. (2004) Cultural clusters and post-industrial city: Towards the remapping of urban cultural policy, *Urban Studies*, 41(3), pp. 507–532.

Peck, J. (2012) Recreative city: Amsterdam, vehicular ideas and the adaptive spaces of creativity policy, *International Journal of Urban and Regional Research*, 36(3), pp. 462–485.

Pine, J. & Gilmore, J. H. (1998) Welcome to the experience economy, *Harvard Business Review*, 76(4), pp. 97–105.

Plaza, B. (2006) The return on investment of the Guggenheim Museum Bilbao, *International Journal of Urban and Regional Research*, 30(2), pp. 452–67.

Poon, J. P. H. & Lai, C. A. (2008) Why are non-profit performing arts organisations successful in mid-sized US cities? *Urban Studies Journal*, 45(11), pp. 2273–2289.

Sacco, P. L. & Crociata, A. (2012) A conceptual regulatory framework for the design and evaluation of complex, participative cultural planning strategies, *International Journal of Urban and Regional Research*. doi: 10.1111/j.1468-2427.2012.01159.x.

Santagata, W. (2002) Cultural districts and property rights and sustainable economic growth, *International Journal of Urban and Regional Research*, 26(1), pp. 9–23.

Sassoon, D. (2006) *The Culture of the Europeans; From 1800 to the Present* (London: Harper Press).

Scott, A. J. (1999) The cultural economy: Geography and the creative field, media, *Culture & Society*, 21, pp. 807–817.

Scott, A. J. (2000) *The Cultural Economies of Cities: Essays on the Geography of Image-producing* (London: Sage).

Scott, A. J. (2004) Cultural-products industries and urban economic development: Prospects for growth and market contestation in global context, *Urban Affairs Review*, 39(4), pp. 461–490.

Scott, A. J. (2008) *Social Economy of the Metropolis: Cognitive-Cultural Capitalism and the Global Resurgence of Cities* (Oxford: Oxford University Press).

Scott, A. J. (2011) Emerging cities of the third wave, *City*, 15(3–4), pp. 289–321.

Steel, C. (2008) *Hungry City; How Food Shapes Our Lives* (London: Chatto & Windus).

Throsby, D. (2010) *The Economics of Cultural Policy* (Cambridge: Cambridge University Press).

Towse, R. (2010) *A Textbook of Cultural Economics* (Cambridge: Cambridge University Press).

Trip, J. J. (2007) *What Makes a City; Planning for "Quality of Place"; The Case of High-Speed Train Station Area Redevelopment* (Amsterdam: IOS Press).

Zukin, S. (1995) *The Cultures of Cities*, (Cambridge, MA: Blackwell Publishers). Available at http://www.iamsterdam.com/en/placestogo/madame-tussauds-amsterdam/012ee768-d585–4cf9-84e8-184a111618d0 (accessed 26 March 2012).

van Zwol, C. (2012) EYE Film Instituut, *NRC Handelsblad*, 2 February.

Index

advertising industry 16
AHRC Creative Economy Hubs 39
Amsterdam Declaration 50
Amsterdam, Netherlands 5; cultural planning 82–97; EYE Film Instituut 83, 87, 91–2, 95; Madame Tussaud's 87, 88
Arts and Humanities Research Council 39

buzz 54, 55

Cameron, David 25
chatter 54, 55
Cities/Capitals of Culture 50; *see also* creative cities
clusters 29, 32, 34, 50, 52, 64, 66, 67, 73
community development 53, 84, 85, 94
competitiveness 4, 8, 23, 24, 34, 50, 53
confrontation matrix 70
consumption milieu 64–5; place qualities 65–7
Coventry, UK 4, 5, 54–9
creative cities 50, 62–81; conceptual background 63–8; confrontation matrix 70; Delft, Netherlands 4, 5, 71–7; place qualities 65–8; policy intervention 1–7, 70–1, 75–7; production/consumption milieu 64–5; SWOT analysis 69–70; *see also* cultural and creative industries
Creative Economy Programme 35
creative industries *see* cultural and creative industries
Creative Industries Knowledge Transfer Network 39
Creative Industries Mapping Document 33
creativity 1–2, 4; *see also entries under creative*
cultural amenities 84, 86, 87–93; iconic projects 91; mass culture 93; niche-oriented 89–90; small scale 90; typology 89; *see also* cultural planning
cultural and creative industries (CCIs) 2, 3, 5, 8–27, 29–30, 33, 34, 36–7, 38–43, 50; Delft, Netherlands 71–7; freelancers in 10–17, 19,

22–4; importance of 8–9; London, UK 8–27; policy intervention 1–7, 33–5, 70–1, 75–7; project-based nature 16; Triple Helix framework 3, 30–43; value chain 71; *see also* creative cities; and specific sectors
cultural planning 4, 82–97; iconic projects 91–2; strategy 85–7, 91; *see also* cultural amenities
culture 1–2, 86, 93–4; *see also entries under cultural*

Delft, Netherlands 4, 5, 71–7; SWOT analysis 73–5
disguised employment 10
doing, using and interacting 51

economic growth 2, 8, 33, 49–51, 60, 84
Economic and Social Research Council 39
enterprising self 14
Environmental and Physical Sciences Research Council 39
European Creative Districts 50
European Innovation Scoreboard 51
European Union 50
EYE Film Instituut (Amsterdam, Netherlands) 83, 87, 91–2, 95

false freelancers 14
festivals 57
filieres 32
firms 10
Florida, Richard 1, 4, 5, 35, 36, 62, 63, 64, 65–6, 77, 86
forced freelancers 14
freelancers/freelancing 3, 8–27; companies hiring 18; definition of 10; false freelancers 14; forced freelancers 14; formal contracts 19–20; importance of 22–4; nature of work 14; number of workers 15; role in creative economy 10–17; true freelancers 14, 23; type of creative work 19; vulnerability of 17, 24–5